Fundamentals of Geometric Dimensioning and Tolerancing

by
Alex Krulikowski

First Edition

 Delmar Publishers Inc.®

NOTICE TO THE READER

Delmar Staff

    Executive Editor: David C. Gordon
    Senior Administrative Editor: Michael A. McDermott
    Project Editor: Carol Micheli
    Production Coordinator: Teresa Luterbach
    Art Coordinator: Michael Nelson
    Design Coordinator: Karen Kunz Kemp

For information, address Delmar Publishers Inc.
2 Computer Drive West, Box 15-015
Albany, New York 12212

Printed in the United States of America
Published simultaneously in Canada
by Nelson Canada,
a division of the Thomson Corporation

10 9 8 7 6 5 4 3 2

Library of Congress Cataloging-in-Publication Data

Krulikowski, Alex
    Fundamentals of geometric dimensioning and tolerancing / by Alex Krulikowski.
        p. cm.
    Includes index
    ISBN 0-8273-4694-8
        1. Engineering drawings--Dimensioning   2. Tolerance (Engineering)
I. Title.
T357.K78 1991   90-23563
604.2 43--dc20   CIP

TABLE OF CONTENTS

CHAPTER		PAGE
1	GEOMETRIC DIMENSIONING	1
2	FORM CONTROLS	33
3	DATUMS	65
4	ORIENTATION CONTROLS	95
5	LOCATION CONTROLS	121
6	RUNOUT CONTROLS	159
7	PROFILE CONTROLS	177
8	GLOSSARY	193
9	CLASS EXERCISES	
	Exercises	199
	APPENDIXES	237
	Feature Control Frame Size Proportions	238
	Tolerance Zone Conversion	239
	Geometric Tolerance Summary Chart	243
	Goemetric Tolerancing Standards Sources	244
	ANSI vs ISO Comparison Chart	245
	ANSWER GUIDE	247
	INDEX	259

FOREWORD

This textbook is part of Effective Training's program on The Fundamentals Of Geometric Dimensioning and Tolerancing. A Geometric Dimensioning and Tolerancing Instructor's Kit is also available. The Instructor's Kit includes a comprehensive Instructor's Guide and Overhead Transparency Masters.

For Information Contact:

Effective Training Inc.
2096 S. Wayne Road
Westland, Michigan 48185

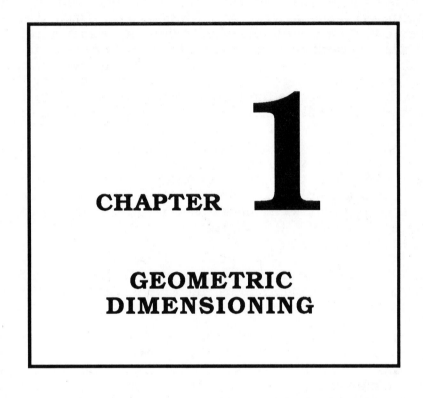

CHAPTER 1

GEOMETRIC DIMENSIONING

Many of industry's problems stem from faulty communications. In today's competitive market, it is not enough to make drawings that can be understood. We must make drawings that cannot possibly be MISUNDERSTOOD.

INTRODUCTION

Studying Geometric Tolerancing is like constructing a building. If you want the building to be strong and last a long time, you must begin by constructing a solid foundation. Likewise, if you want to build an understanding of Geometric Dimensioning and Tolerancing that will be strong and lasting, you must begin by establishing a solid understanding of the fundamentals of the language. By studying the terminology in this chapter, you will be prepared to learn and use the concepts in the following chapters.

GEOMETRIC DIMENSIONING AND TOLERANCING:
A BRIEF HISTORY

As long as people have made things, they have used measurements, drawing methods and drawings.

Drawings existed as far back as 6,000 B.C. Back then, a unit of measure in the Nile and Chaldean civilizations was a "royal cubit." For a couple thousand years it fluctuated anywhere from 18 to 19 inches in length. But around 4,000 B.C. the royal cubit was standardized at 18.24 inches. This set a pattern that has held true for nearly 6,000 years since: as long as there are measurements, drawing methods and drawings, there will be controversies, committees, and standards.

Manufacturing as we know it began with the Industrial Revolution in the 1800's. There were, of course, drawings, but these drawings were very different from the ones we use today. A typical drawing from the 1800's was a neatly inked, multi-viewed artistic masterpiece which portrayed the part with almost pictorial precision. Occasionally the designer would write in a dimension, but generally such things were considered unneccessary.

Why? Because the manufacturing process was very different then. There were no assembly lines, no widely dispersed departments or corporate units scattered across the nation or even world-wide.

Back in those days, manufacturing was a cottage industry employing craftsmen who did it all, from parts fabrication to final assembly, and who passed their hard-won skills down from generation to generation. To these craftsmen, there was no such thing as variation. Nothing less than perfection was good enough.

Of course, there was variation; but back then the measuring instruments were not precise enough to identify it. When misfits and assembly problems occurred, which they routinely did, the craftsmen would simply cut-and-try, file-and-fit till the assembly worked perfectly. The total process was conducted under one roof and communication among craftsmen was immediate and constant: "Keep that on the high side." "That edge has plenty of clearance." "That fit is OK now."

You can see that manufacturing back then was a quality process, but also slow, laborious and consequently quite an expensive one.

The advent of the assembly line and other improved technologies revolutionized manufacturing. The assembly line created specialists to take the place of craftsmen, and these people did not have the time or the skills for "file-and-fit."

Improved methods of measurement also helped to do away with the myth of "perfection." Now, engineers understood that variation is unavoidable. Moreover, in every dimension of every part in every assembly, some variation is acceptable without impairing the function of the assembly -- as long as the limit of that variation -- the "tolerance" -- is identified, understood and controlled. This led to the development of the plus/minus or coordinate system of tolerancing, and the logical place to record these tolerances and other information was the engineering or design drawing.

With this development, drawings became more than just pretty pictures of parts; they became the main means of communication among manufacturing departments which were, increasingly, less centralized, more specialized, and subject to stricter demands.

Engineering Drawing Standards

To improve the quality of drawings, an effort was made to standardize them. In 1935, after years of discussion, the American Standards Association published the first recognized standard for drawings, "American Drawing and Drafting Room Practices." Of its 18 short pages, just five discussed dimensioning; tolerancing was covered in just two paragraphs.

It was a start, but its deficiencies became obvious with the start of World War II. In Britain, wartime production was seriously hampered by high scrap rates due to parts that would not assemble properly. The British determined that this was caused by weaknesses in the plus/minus system of coordinate dimensioning -- and, more critically, by the absence of full and complete information on engineering drawings.

Driven by the demands of war, the British innovated and standardized. Stanley Parker of the Royal Torpedo Factory in Alexandria, Scotland, created a positional tolerancing system calling for round (rather than square) tolerance zones. The British went on to publish a set of pioneering drawing standards in 1944 and, in 1948, published "Dimensional Analysis of Engineering Design." This was the first comprehensive standard using basic concepts of true position dimensioning.

GD & T IN THE UNITED STATES

In the U.S., Chevrolet published a Draftsman's Handbook in 1940, the first publication with any significant discussion of position tolerancing. In 1945 the U.S. Army published its Ordinance Manual on Dimensioning and Tolerancing, which introduced the use of symbols (rather than notes) for specifying form and positioning tolerances.

Even so, the second edition of the American Standard Association's "American Standard Drawing and Drafting Room Practice," published in 1946, made minimal mention of tolerancing. That same year, however, the Society of Automotive Engineers (SAE) expanded coverage of dimensioning practices as applied in the aircraft industry in its "SAE Aeronautical Drafting Manual." An automotive version of this standard was published in 1952.

In 1949, the U.S. military followed the lead of the British by publishing its first standard for dimensioning and tolerancing, known as MIL-STD-8. Its successor, MIL-STD-8A, published in 1953, authorized 7 basic drawing symbols and introduced a methodology of functional dimensioning.

Now there were three different groups in the U.S. publishing standards for drawings: the ASA, the SAE and the military. This led to years of turmoil over the inconsistencies among the standards, and also to slow but measured progress in uniting those standards.

In 1957, the ASA approved the first American standard devoted to dimensioning and tolerancing, in coordination with the British and Canadians; the 1959 MIL-STD-8B brought the military standards closer to ASA and SAE standards; and in 1966, after years of debate, the first united standard was published by the American National Standards Institute (ANSI), successor to ASA: known as ANSI Y14.5. This first standard was updated in 1973 to replace notes with symbols in all tolerancing, and the current standard was published in 1982. An updated ANSI standard is scheduled for publication in 1993.

Geometric Dimensioning and Tolerancing is now in use by 70-80% of all major companies in the United States and is the recognized standard for military contracting.

WHAT IS GEOMETRIC DIMENSIONING AND TOLERANCING

Geometric dimensioning is one of three types of dimensions used on engineering drawings. Figure 1-1 shows how geometric dimensioning fits into the total subject of dimensioning engineering drawings.

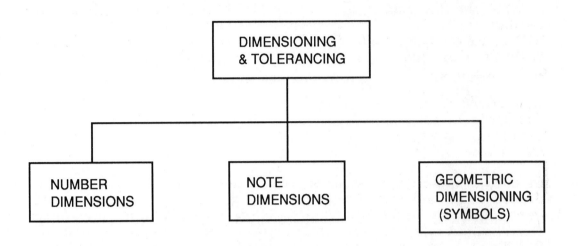

FIGURE 1-1 DIMENSIONING TREE

More specifically, *Geometric Dimensioning and Tolerancing* (G.D.&T.) is a dual purpose system. First it is a set of standard symbols which are used to define part features and their tolerance zones. The symbols and their interpretation are documented by the American National Standards Institute Dimensioning Standard, (ANSI Y14.5M-1982). Second, and of equal importance, geometric dimensioning and tolerancing is a philosophy of defining a part based on how it functions. Geometric Dimensioning readily captures the design intent by providing the designer with better tools to say what he means. Therefore, manufacturing or inspection can understand more clearly the design requirements of the part.

An important concept to remember about geometric dimensioning is that the dimensions on a drawing define what size and shape the part must be to function as the design intended. This dimensioning philosophy is a powerful design tool. It can make a substantial difference in product costs.

Geometric dimensioning and tolerancing, as a dimensioning standard to aid communications, and as a design philosophy to provide the most liberal tolerances, can add up to a substantial savings in a company's operating expenses.

ADVANTAGES

The military, the automotive industry, and many others have been using G.D.&T. for over forty years. One reason this subject has become so popular is simply because its use saves money. The following list highlights a few ways that G.D.&T. can be beneficial to your company.

• Improves communications

G.D.&T can provide uniformity in drawing specifications and interpretation, thereby reducing controversy, guesswork and assumptions. Design, production, and inspection all work to the same language.

[handwritten margin note: uniformity in drawing specs, interpretation for Design, production & inspection.]

• Better product designs

The use of G.D.&T. can improve your product designs. First, by providing designers with tools to "say what they mean", second, by establishing a dimensioning philosophy based on part function. This philosophy, called functional dimensioning, studies product function in the design stage and establishes part tolerances based upon functional requirements.

[handwritten margin note: enables part tolerances to be defined based upon the function of part.]

• Production tolerances increased

There are two ways tolerances are increased through the use of G.D.&T. First, under certain conditions, G.D.&T. provides "bonus" or extra tolerance for manufacturing. This additional tolerance can make a significant savings in production costs. Second, by the use of functional dimensioning, the tolerances are assigned to the part based upon its functional requirements. This often results in a larger tolerance for manufacturing. It eliminates the approach that designers copy existing tolerances, or assign tight tolerances, because they don't know how to determine a reasonable (functional) tolerance.

[handwritten margin note: prodtn costs lowered due to bonus tolerance during Manf]

DISADVANTAGES

There are some shortcomings related to the use of G.D.&T. One is the lack of training available. Currently, few colleges and trade schools offer instruction in this subject. In industry, there are a few courses available. Another shortcoming is the large number of bad examples of G.D.&T. on drawings today. There are literally thousands of drawings in industry today that have uninterpretable or incomplete dimensioning specifications.

This makes it extremely difficult, if not impossible, for drawing users to correctly interpret drawings when there is no correct interpretation. This leads to much confusion. Usually G.D.&T. is blamed, when really the confusion exists because the dimensioning was incorrectly applied.

FUNCTIONAL DIMENSIONING

Functional Dimensioning is a philosophy of dimensioning and tolerancing a part based on how it functions. When functionally dimensioning a part, the designer performs a functional analysis. A *Functional Analysis* is a process where a designer identifies the functions of a part and uses this information to establish the actual part dimensions and tolerances. Functional dimensioning and functional analysis are very powerful design tools. For a designer to become proficient in the use of functional dimensioning it requires several years of concentrated effort. The benefits to the individual, and to the company, reward the effort many times over. Some of the benefits are listed below.

- The designer will develop an objective design philosophy.

- The designer will develop a true understanding of how each part in a design functions.

- Potential product problems will be identified in the design stage.

- An objective method for evaluating change requests will be established.

- Larger tolerances for manufacturing. Tolerances will be based on the "maximum allowable tolerance that will not adversely affect the product function".

- Promote better communications between design and development departments.

- Fewer change requests. In most cases, part tolerances will already be at their maximum value.

DEFINITIONS

Throughout this book, the terms "feature" and "feature-of-size" are used extensively. It is important to understand the meaning and use of these terms. A *feature* is a general term applied to a physical portion of a part, such as a surface, hole, or slot. A *feature-of-size* is one cylindrical or spherical surface or a set of parallel surfaces, each of which is associated with a size dimension.*

* These definitions are paraphrased from ANSI Y14.5-M-1982, SECTION 1-3.

A *Location Dimension* is a dimension which locates the centerline or centerplane of a part feature relative to another part feature, centerline, or centerplane.

Examples of feature, feature-of-size, and location dimensions are shown in Figure 1-2. The chart shows that letters "A" thru "J" represent either a feature or a feature-of-size. The remaining letters, "K,L,M" represent location dimensions.

Location Dimension
is one in which (locates the centerline)
centerplane of a part feature
relative to another part
feature, center line /
center plane.

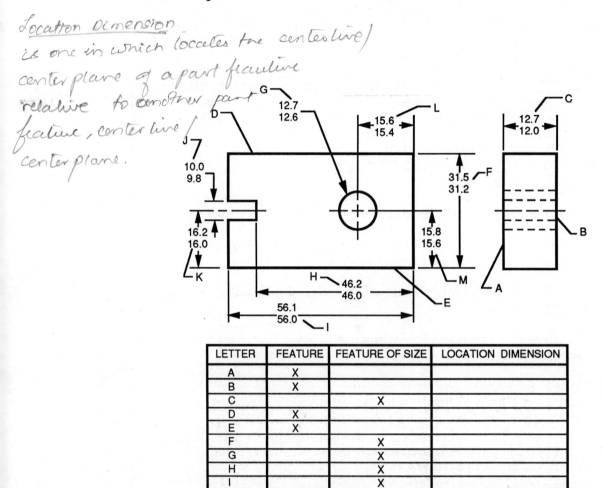

LETTER	FEATURE	FEATURE OF SIZE	LOCATION DIMENSION
A	X		
B	X		
C		X	
D	X		
E	X		
F		X	
G		X	
H		X	
I		X	
J		X	
K			X
L			X
M			X

FIGURE 1-2
IDENTIFICATION OF FEATURES AND FEATURES OF SIZE

When referring to a feature-of-size at one of its extremes, there are terms which are commonly used. It is important to understand the definition of these terms because they are used extensively in this workbook.

When a feature-of-size contains the most amount of material, it is in its *Maximum Material Condition* (MMC). For example, when the pin diameter in Figure 1-3 is at 12.2, the part contains the most amount of material, therefore, it is at maximum material condition. An internal feature-of-size can also have a maximum material condition. When the hole in Figure 1-4 is at 10.0, the part contains the most amount of material, therefore it is at maximum material condition.

REMEMBER

An external feature-of-size (e.g. a shaft) is at MMC when it is at its largest size limit.

An internal feature-of-size (e.g. a hole) is at MMC when it is at its smallest size limit.

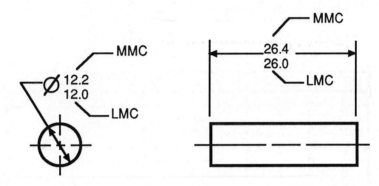

FIGURE 1-3 MMC & LMC OF EXTERNAL FEATURES-OF-SIZE

When a feature-of-size contains the minimum amount of material, it is in its *least material condition* (LMC). For example, when the pin diameter in Figure 1-3 is at 12.0, the part contains the least amount of material; therefore it is at least material condition. Once again, this concept also applies to internal features-of-size. When the hole in Figure 1-4 is at 10.5, the part contains the least amount of material, therefore it is at least material condition.

REMEMBER

An external feature-of-size (e.g. a shaft) is at LMC when it is at its smallest size limit.

An internal feature-of-size (e.g. a hole) is at LMC when it is at its largest size limit.

Another condition that must be covered is how to refer to a feature-of-size which is not at either extreme, but at whatever size it happens to be on a particular part. The term to describe this condition is regardless of feature size. *Regardless Of Feature Size* (RFS) is when a geometric tolerance (or datum) applies independent of the feature size. The geometric tolerance is limited to the stated amount regardless of the size of the feature-of-size.

(RFS) Geometric tolerance applies independent of feature size.

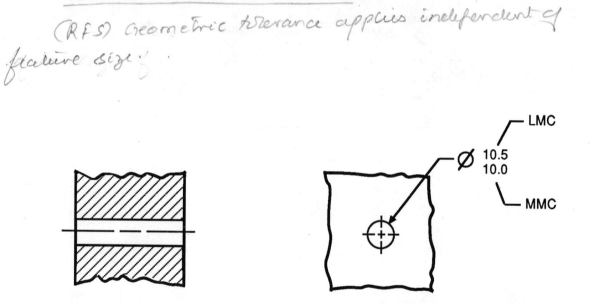

FIGURE 1-4 MMC & LMC OF INTERNAL FEATURE-OF-SIZE

GEOMETRIC CHARACTERISTIC SYMBOLS

There are thirteen geometric characteristic symbols used in the language of G.D.&T. They are shown in Figure 1-5. They are divided into five separate categories: form, orientation, location, runout, and profile. Detailed explanations of each symbol are in the following chapters. See the appendix for size proportions of geometric tolerance symbols.

CATEGORY	CHARACTERISTIC	SYMBOL	DATUM * REFERENCS
FORM	FLATNESS	▱	NEVER USES A DATUM REFERENCE
	STRAIGHTNESS	—	
	CIRCULARITY	○	
	CYLINDRICITY	⌀	
ORIENTATION	PERPENDICULARITY	⊥ /	ALWAYS USES A DATUM REFERENCE
	ANGULARITY	∠ /	
	PARALLELISM	//	
LOCATION	POSITION	⊕ /	
	CONCENTRICITY	◎	
RUNOUT	CIRCULAR RUNOUT	↗	
	TOTAL RUNOUT	↗↗	
PROFILE	PROFILE OF A LINE	⌒	MAY USE A DATUM REFERENCE
	PROFILE OF A SURFACE	⌓ /	

FIGURE 1-5 GEOMETRIC SYMBOLS

* Datums are reference planes used for making part measurement. Datums are explained in Chapter 3.

MODIFYING SYMBOLS

In addition to the geometric characteristic symbols, there are five modifying symbols used in G.D.&T. These modifiers and their symbols are listed in Figure 1-6. The first three modifiers, MMC, LMC, and RFS, have been introduced earlier in this chapter. The fourth modifier, projected tolerance zone, is explained in chapter five. The last modifier, diameter, is represented on drawings by the symbol shown in Figure 1-6.

TERM	ABBREVIATION	SYMBOL
MAXIMUM MATERIAL CONDITION	MMC	Ⓜ
LEAST MATERIAL CONDITION	LMC	Ⓛ
REGARDLESS OF FEATURE SIZE	RFS	Ⓢ
PROJECTED TOLERANCE ZONE		Ⓟ
DIAMETER	DIA	⌀

FIGURE 1-6 MODIFIERS

Based on ANSI Y14.5M-1982. Reproduced by permission of the American Society of Mechanical Engineers.

FEATURE CONTROL FRAME

Geometric tolerances and modifiers are applied to drawings with the use of a feature control frame. A *feature control frame* is a rectangle which is divided into compartments within which the geometric characteristic symbol, tolerance value, modifiers, and datum references are placed. The parts of a feature control frame are shown in Figure 1-7. The placement of feature control frames on a drawing is shown in Figure 1-8. See the appendix for size proportions of feature control frames.

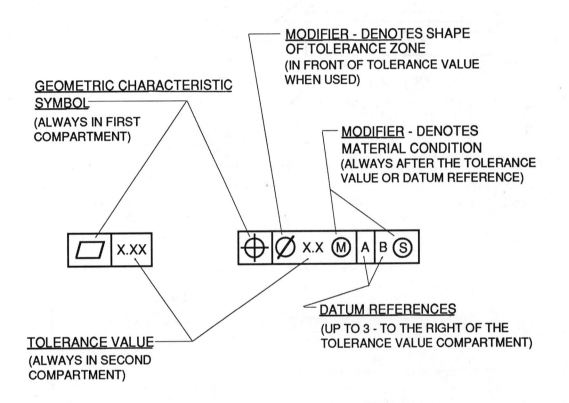

FIGURE 1-7 FEATURE CONTROL FRAMES

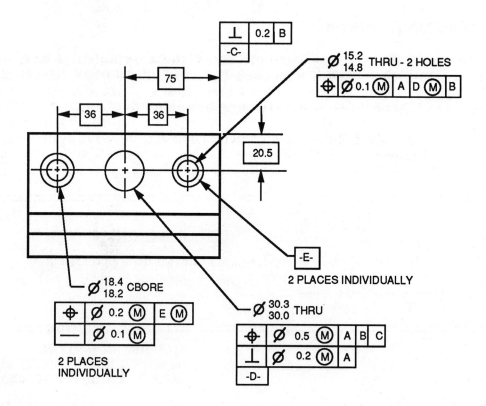

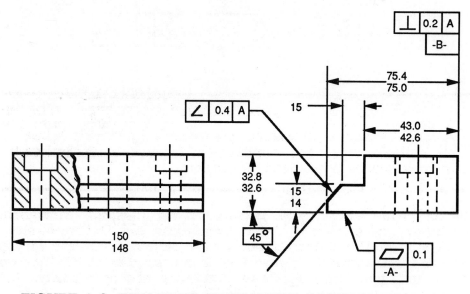

**FIGURE 1-8 EXAMPLES OF FEATURE CONTROL FRAME
PLACEMENT ON A DRAWING**

FUNDAMENTAL RULES

In geometric tolerancing, there are three fundamental rules that are used extensively. These rules are the foundation upon which the system of geometric tolerancing is based. It is extremely important to understand these rules when studying geometric tolerancing.*

The first rule, called Rule #1, is shown below. Examples of Rule #1 are shown in Figure 1-9.

RULE #1   (Envelope rule)

For features-of-size, where only a size dimension is specified, the surfaces shall not extend beyond a boundary (envelope) of PERFECT FORM AT MMC.

AT MMC PERFECT FORM

AT LLC BONUS TOL.

INTERPRETATION

10.0±0.2

9.8

10.2

BONUS OF 0.4

REMEMBER

Rule #1 applies to all features-of-size on a drawing. It is like an invisible control that applies to all features-of-size unless it is overriden by a geometric tolerance.

Perfect "form" in Rule #1 means perfect flatness, straightness, roundness, and cylindricity. In other words, if a feature-of-size was produced at MMC (all over) it would have perfect form. Rule #1 applies to the features-of-size in Figure 1-9. If produced at MMC, they would have to have perfect straightness, circularity, and cylindricity. If they are produced between MMC and LMC, let's say 10.7, then a form error equal to the amount of departure from MMC (10.8-10.7 = 0.1) would be permissible. If a feature-of-size is produced at LMC, a form error equal to the amount of departure (0.2) would be permissible. If this sounds confusing, don't panic. In Chapter 2, the relationship of form controls and Rule#1 are fully explained.

* The rules shown here are paraphrased from ANSI-Y14.5M-1982.

RULE #1 APPLIED TO AN EXTERNAL FEATURE-OF-SIZE

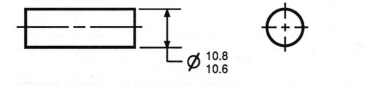

Ø 10.8
10.6

INTERPRETATION PER RULE #1

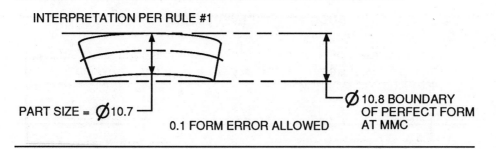

PART SIZE = Ø 10.7

0.1 FORM ERROR ALLOWED

Ø 10.8 BOUNDARY
OF PERFECT FORM
AT MMC

RULE #1 APPLIED TO AN INTERNAL FEATURE-OF-SIZE

Ø 14.6
14.0

INTERPRETATION PER RULE #1

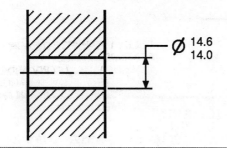

Ø 14.0 BOUNDARY OF
PERFECT FORM AT MMC

HOLE SIZE = Ø 14.6

0.6 FORM ERROR ALLOWED

FIGURE 1-9 EXAMPLES OF RULE #1

Inter Relation- tion feature of size must the controlled big Location/ orientation.

When using Rule #1, it is important to understand exactly what it applies to. Rule #1 applies to individual features-of-size only. Rule #1 does not apply to the interrelationship of features-of-size. Features-of-size shown perpendicular or coaxial to each other must be controlled for location or orientation to avoid incomplete drawing specifications. Figure 1-10 shows an example of Rule #1 applied to the features-of-size which make a box.

REMEMBER

Rule #1 applies to individual feature-of-size only.

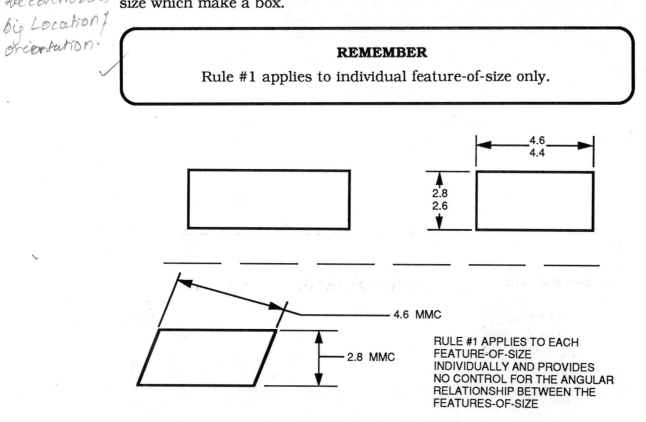

RULE #1 APPLIES TO EACH FEATURE-OF-SIZE INDIVIDUALLY AND PROVIDES NO CONTROL FOR THE ANGULAR RELATIONSHIP BETWEEN THE FEATURES-OF-SIZE

FIGURE 1-10 RULE #1

There are a few exceptions to Rule #1 that you should be aware of. First, Rule #1 does not apply to features-of-size that are subject to free state variation. That is, parts that are not rigid when held in your hand. For the treatment of such parts, see ANSI Y14.5M-1982 section 6.8. The second exception is that Rule #1 does not apply to stock sizes such as bar stock, tubing, sheets, or structural shapes. Paragraph 2.7.1.3 in ANSI-Y14.5M-1982 explains why. It states ". . . Standards for these items govern the surfaces that remain in the "as furnished" condition on the finished part."

REMEMBER

There are two exceptions to rule #1; nonrigid parts and stock sizes.

Rule #1 and the feature-of-size dimension to which it is applied are related to each other in a unique way. Rule #1 states that a feature-of-size must have "perfect" form if the feature-of-size is at MMC. This means that for the part in Figure 1-11, if the 2.6-2.8 dimension was at MMC, the top and bottom of the block would have to be perfectly flat. If the part was less than MMC of the dimension, a form error (i.e. flatness of the surfaces) equal to the amount of departure, would be permissible. If a part were produced at LMC, the form error (i.e. flatness of the surfaces) equal to the amount of the departure would be permissible. This concept will be developed further in Chapter 2.

REMEMBER

Rule #1 and its associated size dimension are interrelated.

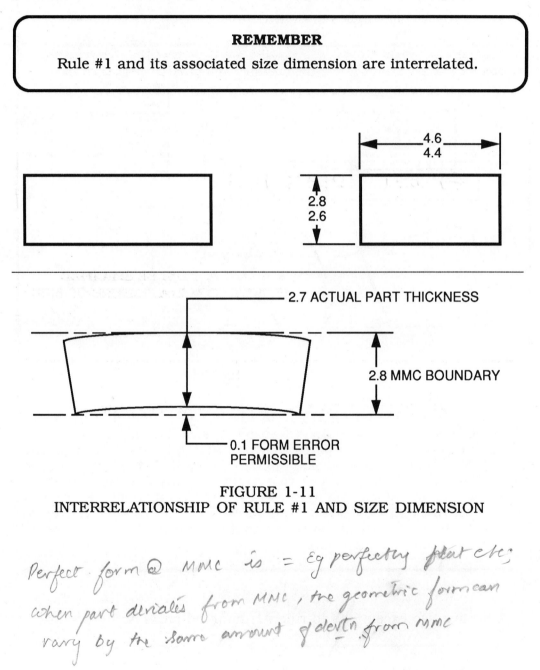

FIGURE 1-11
INTERRELATIONSHIP OF RULE #1 AND SIZE DIMENSION

Perfect form @ MMC is = eg perfectly flat etc;
when part deviates from MMC, the geometric form can
vary by the same amount of devtn from MMC

18

The second and third fundamental rules are simply conventions for expressing geometric tolerances in feature control frames. Figures 1-12 and 1-13 show Rule #2 and Rule #3 respectively. Their use will be explained in the following chapters.

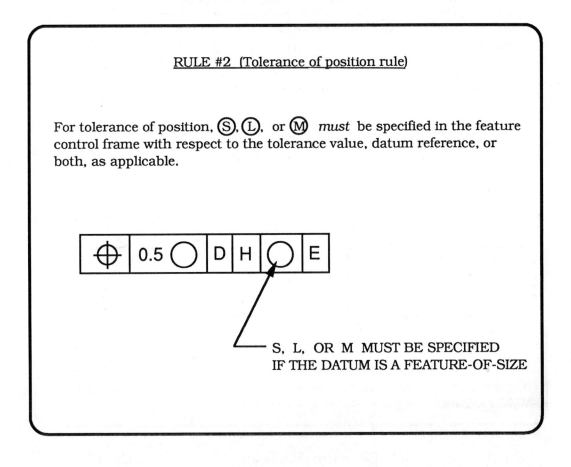

FIGURE 1-12 TOLERANCE OF POSITION RULE

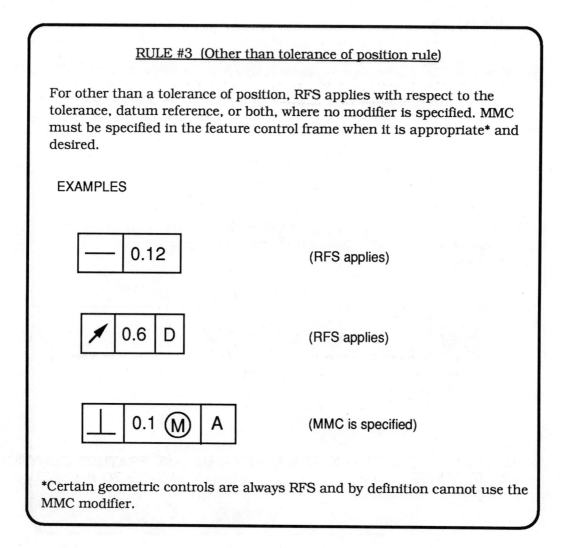

RULE #3 (Other than tolerance of position rule)

For other than a tolerance of position, RFS applies with respect to the tolerance, datum reference, or both, where no modifier is specified. MMC must be specified in the feature control frame when it is appropriate* and desired.

EXAMPLES

| — | 0.12 | (RFS applies)

| ⌗ | 0.6 | D | (RFS applies)

| ⊥ | 0.1 Ⓜ | A | (MMC is specified)

*Certain geometric controls are always RFS and by definition cannot use the MMC modifier.

FIGURE 1-13 OTHER THAN TOLERANCE OF POSITION RULE

APPLICATIONS

A feature control frame can be applied to either a feature or a feature-of-size. It is the location of the feature control frame on the drawing that indicates if it applies to a feature or a feature-of-size. In Figure 1-14 the feature control frame is located so that it applies to the surface of the pin (it is not associated with any feature-of-size dimensions). Therefore, it applies to a feature.

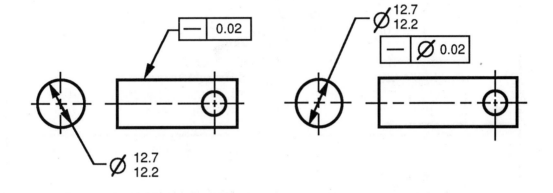

FIGURE 1-14 FEATURE CONTROL FRAME APPLIED TO A FEATURE

FIGURE 1-15 FEATURE CONTROL FRAME APPLIED TO A FEATURE-OF-SIZE

In Figure 1-15, the location of the feature control frame is directly beneath the dimension, indicating that it applies to the dimension of the diameter. The dimension is a feature-of-size, the control applies to the axis of the diameter. In this type of application the feature control frame applies to a feature-of-size.

One important distinction between a symbol applied to a feature versus one applied to a feature-of-size is its effect on Rule #1. Symbols applied to a feature-of-size override Rule #1 and symbols applied to a feature do not override Rule #1.

REMEMBER

The location of a feature control frame determines if it applies to a feature or a feature-of-size.

Only feature control frames which apply to a feature-of-size can override rule #1.

BASIC DIMENSIONS

A *Basic Dimension* is a numerical value used to describe theoretically exact characteristics of a feature or datum target. It is the basis from which permissible variations are established by tolerances on other dimensions, in notes or in a feature control symbol. Previously used "gage" dimensions (untoleranced dimensions used to establish gaging points, lines or planes) are often specified as basic dimensions.

If basic dimensions are used to define part features, geometric tolerances must be added to specify how much tolerance is permissible from the exact location described by the basic dimensions. If basic dimensions are used to define gage dimensions, such as datum targets*, then no geometric tolerances are used to define tolerance for the basic dimension. In this case gage makers tolerances (a very small tolerance when compared to product tolerances) will apply.

Basic dimensions are usually specified in one of three ways; enclosing the numerical value in a rectangle, placing the word "basic" after the dimension, or by the use of a general note. Figure 1-16 illustrates these methods.

REMEMBER

Basic Dimensions which define part characteristics must have a geometric tolerance to specify a tolerance for the part feature.

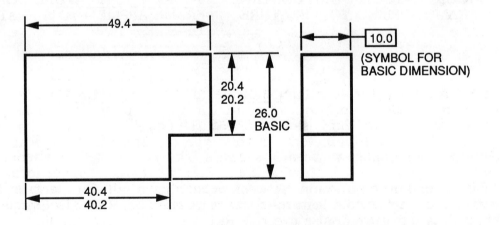

UNLESS OTHERWISE SPECIFIED
ALL UNTOLERANCED DIMENSIONS
ARE BASIC

FIGURE 1-16 BASIC DIMENSIONS

---

* Datum targets are gage points. See the datum target section in Chapter 3 for a full explanation.

BONUS TOLERANCE

Whenever a geometric tolerance is applied to a feature-of-size, and it contains an MMC modifier in the tolerance portion of the feature control frame, a bonus tolerance is possible. When the MMC modifier is used in this fashion, it means that the stated tolerance applies when the feature-of-size is at its maximum material condition. When the actual feature-of-size departs from MMC, an increase in the stated tolerance, equal to the amount of the departure, is permitted. This increase or - extra tolerance - is called the bonus tolerance. Figure 1-17 shows how bonus tolerance is calculated for a straightness application.

*[handwritten margin note: * Bonus Tolerance possible only if F. of size has a MMC modifier]*

REMEMBER

The bonus tolerance value comes from the feature-of-size tolerance. It is equal to the amount that the feature-of-size departs from MMC.

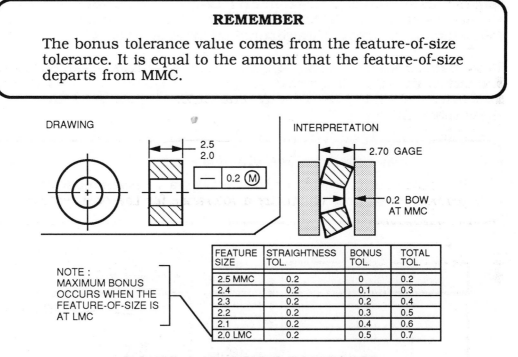

FEATURE SIZE	STRAIGHTNESS TOL.	BONUS TOL.	TOTAL TOL.
2.5 MMC	0.2	0	0.2
2.4	0.2	0.1	0.3
2.3	0.2	0.2	0.4
2.2	0.2	0.3	0.5
2.1	0.2	0.4	0.6
2.0 LMC	0.2	0.5	0.7

NOTE :
MAXIMUM BONUS OCCURS WHEN THE FEATURE-OF-SIZE IS AT LMC

FIGURE 1-17 BONUS TOLERANCE

Another case that may be considered a bonus tolerance is whenever Rule #1 applies to a feature-of-size. Rule #1 states "perfect form at MMC" but, when the feature-of-size departs from MMC, a form error equal to the amount of the departure, is permissible. See discussion of Rule #1 in this chapter.

REMEMBER

There are two conditions where a bonus tolerance is available:

- Whenever Rule #1 applies.
- Whenever a geometric tolerance is applied to a feature-of-size at MMC.

VIRTUAL CONDITION

When analyzing parts that assemble with other parts, or when designing gages, it is essential to be able to calculate a theoretical extreme boundary for part features. *Virtual Condition* is the theoretical extreme boundary of a feature-of-size generated by the collective effects of MMC and any applicable geometric tolerances.

All features-of-size have a virtual condition. When a feature-of-size has no geometric tolerances applied to it, its virtual condition is equal to its MMC plus the effects of Rule #1. This condition is illustrated in Figure 1-18. When a feature-of-size has a geometric tolerance applied to it which overrides Rule #1, then its effects must be considered in determining the virtual condition. An example is shown on page 26.

The Virtual Condition concept is used by three groups:

• Product Designers - To calculate extreme conditions for analyzing mating parts.

• Inspectors - To determine extreme conditions for open inspection set-up.

• Gage Designers - To calculate gage dimensions.

V. Condition =   *MMC + Geometric tolerance if applicable*
@ No geometric   / *MMC + Rule #1*
                  Any deviation from perfect form

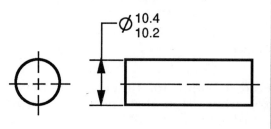

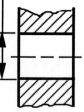

MMC = 10.4	MMC = 10.4
VIRTUAL CONDITION = 10.4 + 0 = 10.4	VIRTUAL CONDITION = 10.4 - 0 = 10.4
EXTERNAL FEATURE-OF-SIZE	INTERNAL FEATURE-OF-SIZE

FIGURE 1-18 VIRTUAL CONDITION AS ESTABLISHED BY RULE #1

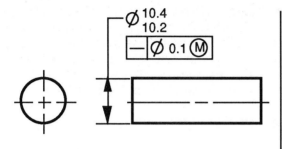
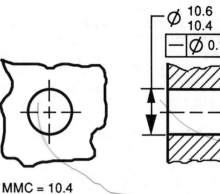

MMC = 10.4

VIRTUAL CONDITION = 10.4 + 0.1 = 10.5

EXTERNAL FEATURE-OF-SIZE

MMC = 10.4

VIRTUAL CONDITION = 10.4 + 0.1 = 10.3

INTERNAL FEATURE-OF-SIZE

FIGURE 1-19   VIRTUAL CONDITION ESTABLISHED USING
GEOMETRIC TOLERANCES

VOCABULARY WORDS

Geometric Dimensioning and Tolerancing   (G.D.&T.)
Functional Dimensioning
Functional Analysis
Feature
Feature-of-size
Location Dimension
Maximum Material Condition
Least Material Condition
Regardless of Feature Size
Rule #1
Rule #2
Rule #3
Feature Control Frame
Basic Dimension
Bonus Tolerance
Virtual Condition

THOUGHT QUESTIONS

Good interpretation
less confusion
prove to be
costly if not mandated

1. List four benefits of having complete drawing specifications.

2. What is the effect of making a part too good? *no bonus tolerance available during assembly*

3. List four possible outcomes of a drawing with dimensions missing.

4. Why is virtual condition important?
   Gage Mount

5. How does geometric tolerancing improve communications on drawings?

PROBLEMS AND QUESTIONS

1. Geometric tolerancing and dimensioning is a dual purpose system. First it *set of std symbols to define part features & their tolerance zone.*

   Second, and of equal importance, it is a *philosophy of define a part based on its functions.*

2. List the five categories of Geometric Characteristic Controls.

FORM	□ — ○ ⌭
ORIENTATION	⊥ // ∠
LOCATION	⊕ ◎
Runout	↗ ↗↗
Profile	⌒ ⌓

3. List three advantages of G.D.&T.

   Improves communication between Manf, Prodtn & Inspctn
   Better product design (based on function of part)
   prodtn costs lowered due to bonus tolerance

4. Functional Dimensioning is *tolerancing based upon the functional of the part design intent*

26

5. What do the symbols in this chapter represent?

    _____

    _____

    _____

    _____

    _____

6. Define Rule #1 _For feature-of-size where only a size_
    dimension is specified, the surfaces shall not extend the boundary
    (envelope) of perfect form @ MMC.

7. Define virtual condition _is the theoretical extreme boundary of_
    a feature-of-size generated by the collective effects of MMC
    & any applicable geometric tolerance.

8. Define basic dimension _is the numerical value used to_
    to theoretically describe the exact characteristics of a feature/
    datum target
    - It is used as a basis from which permissible variations 'r'
    established by tolerance on other dimensions.

9. Describe bonus tolerance.

    when actual feature-of-size departs from
    MMC any deviation from MMC is
    defined as bonus tolerance.

10. What is the virtual condition of the pin shown below?

    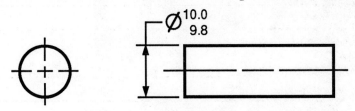

    Ø 10.0
      9.8

    VIRTUAL CONDITION IS ___10.0___

11. In the chart below, place an "x" in the column to indicate if the letter in the drawing is identifying a feature, feature of size, or a location dimension.

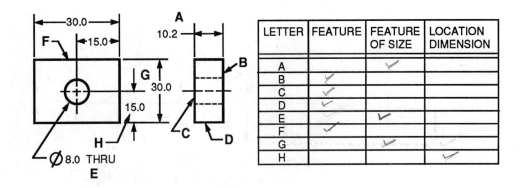

LETTER	FEATURE	FEATURE OF SIZE	LOCATION DIMENSION
A		x	
B	x		
C	x		
D	x		
E		x	
F	x		
G		x	x
H			x

12. In the chart below, place an "x" in the column to indicate if the letter in the drawing is identifying a feature, feature of size, or a location dimension.

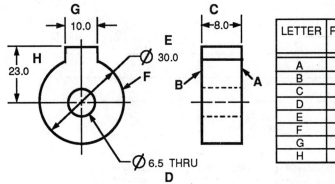

LETTER	FEATURE	FEATURE OF SIZE	LOCATION DIMENSION
A	x		
B	x		
C		x	
D		x	
E		x	
F	x		
G		x	
H			x

28

13. In the chart below, fill in the value of the MMC and LMC for each dimension.

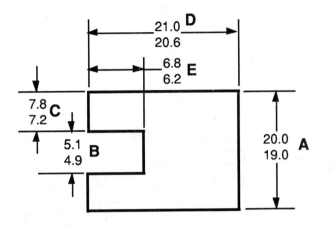

LETTER	MMC	LMC
A	20.0	19.0
B	4.9	5.1
C	7.8	7.2
D	21.0	20.6
E	6.2	6.8

14. In the chart below, fill in the value of the MMC and LMC for each dimension.

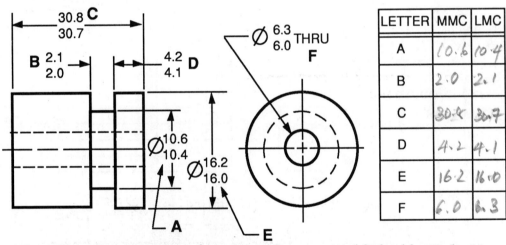

LETTER	MMC	LMC
A	10.6	10.4
B	2.0	2.1
C	30.8	30.7
D	4.2	4.1
E	16.2	16.0
F	6.0	6.3

15. Fill in the maximum form boundary as established by Rule #1.

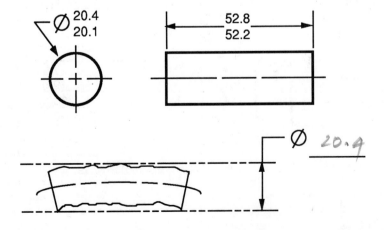

Ø 20.4

29

16. Fill in the minimum form boundary as established by Rule #1.

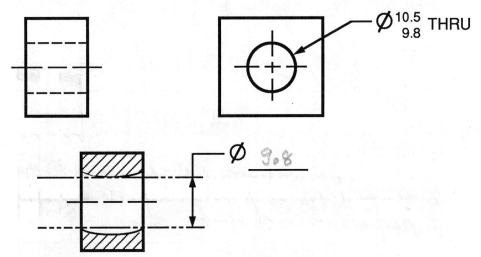

$\emptyset$ $\underline{9.8}$

17. In the chart below, fill in the bonus tolerance value.

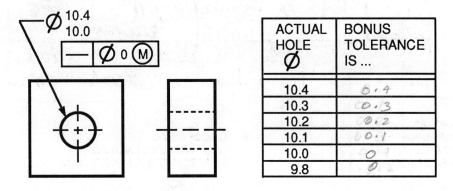

ACTUAL HOLE $\emptyset$	BONUS TOLERANCE IS ...
10.4	0.4
10.3	0.3
10.2	0.2
10.1	0.1
10.0	0
9.8	0

18. On the part above, the virtual condition is ___9.8___.

19. In the chart below, fill in the bonus tolerance value.

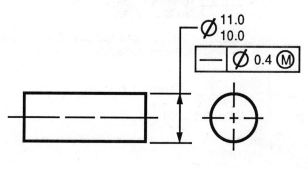

ACTUAL PIN $\emptyset$	BONUS TOLERANCE IS ...
11.0	0
10.8	0.2
10.6	0.4
10.4	0.6
10.2	0.8
10.0	1.0

20. On the part above, the virtual condition is ___11.4___.

ACROSS

1. GOES ON CEREAL
3. DOWN
4. REFERENCE PLANE
7. PART OF WHEEL
9. BAD PART
10. _____ DIMENSIONING
15. VERBAL
16. STAMINA
17. MAXIMUM MATERIAL CONDITION (ABBR)
18. NUMBER OF RULES
20. TYPE OF TOLERANCE
22. _____ CONDITION
24. EXTRA TOLERANCE
25. PRIOR TO: PREFIX
26. ONE WHO MAKES DRAWINGS
29. REGARDLESS OF FEATURE SIZE (ABBR)
30. GEOMETRIC TOLERANCING
    _____ COMMUNICATIONS
31. MACHINE PART
33. FROZEN
37. MANAGER
39. TYPE OF RUNOUT CONTROL
41. PERFECT FORM AT MMC
42. DOWNCAST
43. MOVING TRUCK
44. DESIGN _____
46. AT ANY TIME
47. IMMEDIATE
48. ORIENTATION TOLERANCE

DOWN

1. BODY TISSUE
2. TECHNICAL BOOK AUTHOR
3. LEAST MATERIAL CONDITION (ABBR)
4. LIONS' HOME
5. UNITED STATES (ABBR)
6. MASTER OF CEREMONIES (ABBR)
8. OUT (OPP)
11. ORIGINAL EQUIPMENT MANUFACTURER (ABBR)
12. SCARF
13. DEPICTS TOLERANCE ZONE SHAPE
14. _____ CONTROL FRAME
19. INTERNAL FEATURE-OF-SIZE
21. FIX
23. AMERICAN NATIONAL STANDARDS INSTITUTE (ABBR)
24. DIMENSION WITH NO TOLERANCE
27. FEATURE
28. NOT HARD
32. _____ TOLERANCE
34. LOCATION CONTROL
35. FORM TOLERANCE
36. _____ ENGINE
38. STAIR
40. DEPEND
41. OLDEN CAR
45. GOLF TERM

CHAPTER 2

FORM CONTROLS

When you can measure what you are speaking about and express it in numbers, you know something about it, when you cannot express it in numbers, your knowledge is of a meager and unsatisfactory kind.

Lord Kelvin 1883

INTRODUCTION

Form controls refine or expand the shape of a feature or feature-of-size whenever the limits established by Rule #1 are not functionally satisfactory or applicable. In this chapter, the basic principles of form control are discussed.

GENERAL INFORMATION

When discussing the form of an object, one is discussing the flatness of its surfaces, the straightness of its line elements, the roundness of a circular section, or how cylindrical the object is. On engineering drawings, these conditions are covered by four symbols. They are:

```
FLATNESS          ⟋ ⟍

STRAIGHTNESS      ——

CIRCULARITY       ○

CYLINDRICITY      ⌀
```

Whenever the boundaries established by tolerances of size, location, and Rule #1 do not supply sufficient control to satisfy part functional requirements, then a form tolerance is applied.

Form controls always apply to single features or features-of-size. Form controls are used to define the shape of a feature in relation to itself. Therefore, form controls never use a datum reference.*

REMEMBER

Form controls never use a datum reference.

In the application of form controls, Rule #3 applies. See Chapter 1, page 20 for description of Rule #3.

* Datums are reference planes used for making part measurements. Datums are explained in Chapter 3.

FLATNESS

When a surface is flat, all of its elements fall into a single true plane. A *flatness tolerance* is the amount which surface elements are permitted to vary from a true plane. A flatness tolerance zone is the distance between two planes. Flatness as well as other form tolerances are measured in relation to their own true counterpart. In the case of flatness, a theoretical plane is established by the three high points of the considered surface, a second plane is parallel to the first, but offset by the flatness tolerance value. All points of the considered surface must lie between these two planes. See Figure 2-1.

> **REMEMBER**
>
> A flatness tolerance zone is the distance between two parallel planes spaced apart a distance equal to the flatness tolerance value.

DRAWING

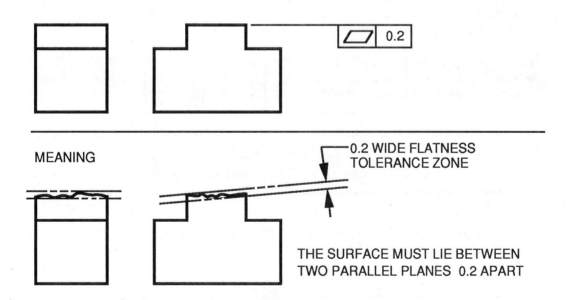

MEANING

0.2 WIDE FLATNESS
TOLERANCE ZONE

THE SURFACE MUST LIE BETWEEN
TWO PARALLEL PLANES 0.2 APART

FIGURE 2-1 FLATNESS TOLERANCE ZONE

Flatness can only be applied to a planar surface. Therefore, it cannot use the MMC or LMC modifier. A geometric control can only use these modifiers when it is applied to a feature-of-size. Also Rule #3 applies.

Modifiers applicable to feature of size only

RULE #1

Whenever Rule #1 applies to a planar feature-of-size, an automatic flatness control exists for both surfaces. This automatic control is a result of the interrelationship of Rule #1 (perfect form at MMC) and the size dimension. When the feature of size is at MMC, both surfaces must be perfectly flat. As the feature-of-size departs from MMC, a flatness error, equal to the amount of departure is allowed. See Figure 2-2.

REMEMBER

Whenever Rule #1 applies to a planar feature-of-size, it provides a flatness control for both surfaces.

DRAWING

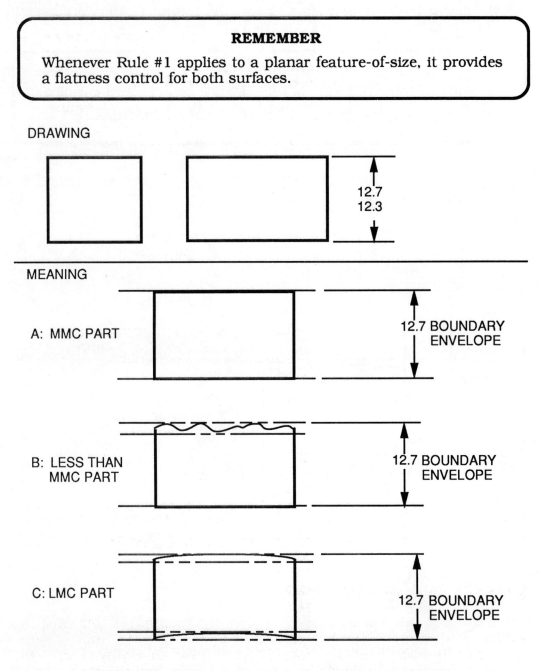

12.7
12.3

MEANING

A: MMC PART — 12.7 BOUNDARY ENVELOPE

B: LESS THAN MMC PART — 12.7 BOUNDARY ENVELOPE

C: LMC PART — 12.7 BOUNDARY ENVELOPE

FIGURE 2-2 RULE #1 AS A FLATNESS CONTROL

APPLICATION

When the automatic flatness control from Rule #1 is not sufficient to satisfy the functional requirements of a part surface - a flatness control may be added. A flatness control never overrides Rule #1 - it refines the maximum allowable flatness error of the surface. Figure 2-3 shows an application of a flatness control. The following statements list several observations which can be made from studying the example in Figure 2-3.

- The flatness control limits the surface flatness only when the part departs from MMC by more than the flatness tolerance value.

- The flatness control does not override rule #1.

- The flatness control does not affect virtual condition.

- The flatness tolerance value should be less than the size tolerance.

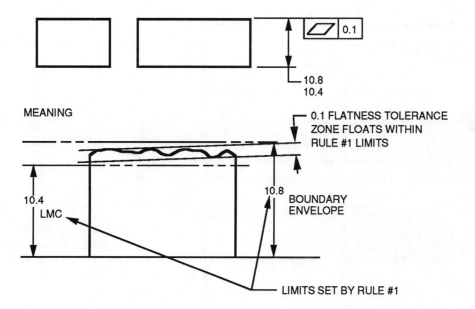

FIGURE 2-3   FLATNESS APPLICATION

INDIRECT FLATNESS CONTROLS

There are a number of geometric controls that can affect flatness of a surface indirectly. The interrelationship of Rule #1 and a size dimension is such a control (see Chapter 1 for an explanation). In certain cases, perpendicularity, parallelism, angularity, total runout, and profile can also control flatness indirectly.

LEGAL SPECIFICATION TEST

For a flatness control to be a legal specification, it must satisfy the following conditions:

- No datum reference may be used in the feature control frame.

- The control must be applied to a planar feature.

- No modifiers may be used.

- The flatness tolerance value specified must be a refinement of any other geometric tolerances that may control the form of the feature (e.g. Rule #1 perpendicularity, parallelism, angularity, total runout, and profile).

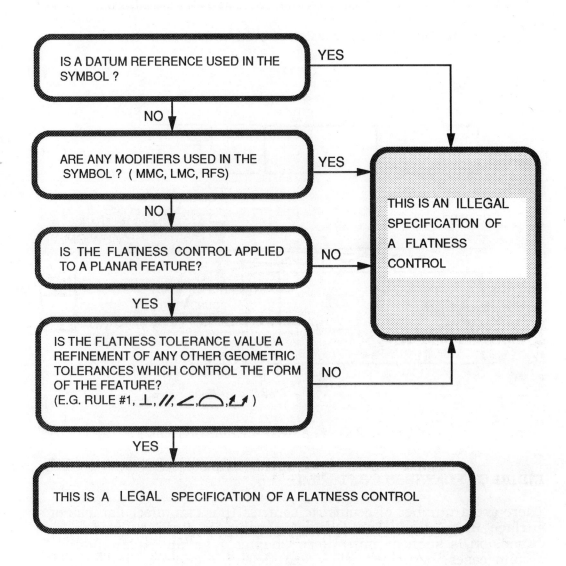

FIGURE 2-4 TEST FOR LEGAL SPECIFICATION OF FLATNESS

STRAIGHTNESS OF FEATURES

Straightness is a condition where each line element of a feature is a theoretical straight line. A *straightness tolerance*, of a feature, is the amount a surface line element is permitted to vary from a theoretically straight line. The shape of a surface element straightness tolerance zone is two parallel lines. The distance between the lines is the tolerance value specified in the feature control frame. The tolerance zone applies in the view that the straightness symbol is shown on the drawing - it offers no control in the other view. See Figure 2-5.

REMEMBER

A straightness tolerance zone for a surface element (feature) is two parallel lines spaced a distance equal to the straightness tolerance value.

DRAWING

MEANING

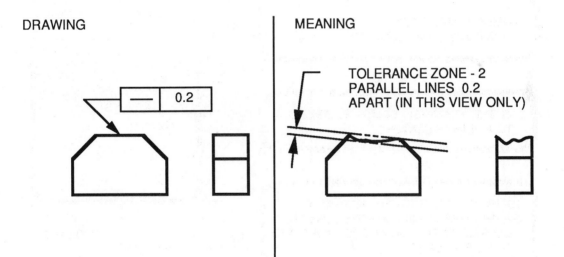

TOLERANCE ZONE - 2
PARALLEL LINES 0.2
APART (IN THIS VIEW ONLY)

FIGURE 2-5 STRAIGHTNESS OF A FEATURE

Straightness of a feature can only be applied to a surface element. Therefore, it cannot use the MMC or LMC modifier. A geometric control can use these modifiers only when it is applied to a feature-of-size. Also Rule #3 applies.

RULE #1

Whenever Rule #1 applies to a planar feature-of-size, an automatic straightness control exists for the surface elements of both surface(s). This automatic control is a result of the interrelationship of Rule #1 (perfect form at MMC) and the size dimension. When the feature-of-size is at MMC, the line elements of the surface(s) must be perfectly straight. As the feature of size departs from MMC, a straightness error equal to the amount of the departure is allowed. See Figure 2-6

> **REMEMBER**
>
> Whenever Rule #1 applies to a planar feature-of-size, it provides a straightness control for the line elements of both surfaces (features).

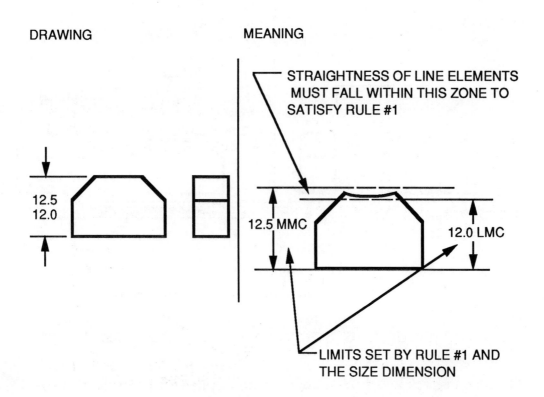

DRAWING   MEANING

STRAIGHTNESS OF LINE ELEMENTS MUST FALL WITHIN THIS ZONE TO SATISFY RULE #1

12.5
12.0

12.5 MMC

12.0 LMC

LIMITS SET BY RULE #1 AND THE SIZE DIMENSION

FIGURE 2-6 RULE #1 AS A STRAIGHTNESS CONTROL

APPLICATION

If the automatic indirect straightness control from Rule #1 is not sufficient to satisfy the functional requirements of a part surface - a straightness control may be added. A straightness control applied to surface elements never overrides Rule #1 - it refines the maximum allowable straightness error of the surface(s). The following statements list several observations which can be made from studying the example in Figure 2-7.

- The straightness control limits the surface element(s) only when the part departs from MMC by more than the straightness tolerance value.

- The straightness control does not override Rule #1.

- The straightness control does not affect the virtual condition.

- The straightness tolerance value should be less than the size tolerance.

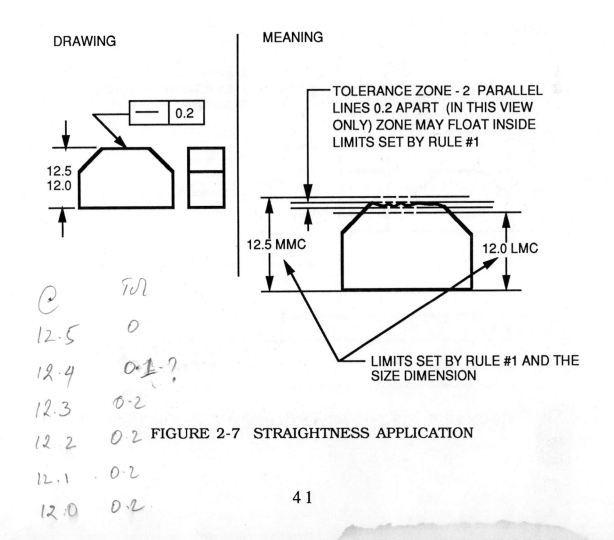

DRAWING

MEANING

TOLERANCE ZONE - 2 PARALLEL LINES 0.2 APART (IN THIS VIEW ONLY) ZONE MAY FLOAT INSIDE LIMITS SET BY RULE #1

12.5
12.0

0.2

12.5 MMC

12.0 LMC

LIMITS SET BY RULE #1 AND THE SIZE DIMENSION

FIGURE 2-7 STRAIGHTNESS APPLICATION

41

LEGAL SPECIFICATION TEST

For a straightness control to be a legal specification, it must satisfy the following conditions:

- No datum reference may be used in the feature control frame.

- The control must be applied to a feature's surface.

- Straightness control should be shown in the view where the line elements being controlled are shown as a line.

- No modifiers may be used.

- The tolerance value specified must be a refinement of any other geometric tolerances that control the form of the feature (e.g. Rule #1, perpendicularity, parallelism, angularity, profile, runout, etc.).

A simple test for verifying if a straightness control (applied to a feature) is legal is shown in Figure 2-8.

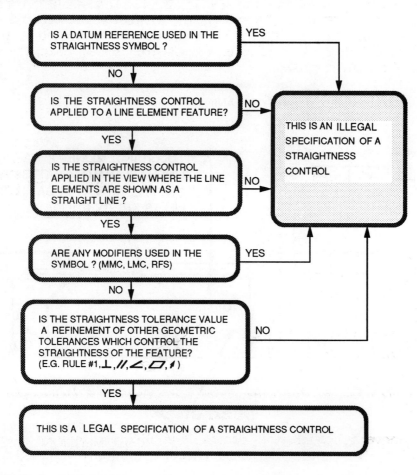

FIGURE 2-8 TEST FOR LEGAL SPECIFICATION OF STRAIGHTNESS
APPLIED TO A FEATURE

STRAIGHTNESS OF FEATURES-OF-SIZE

Straightness is the only form tolerance that can be applied to either a feature or a feature-of-size. A straightness control has a different tolerance zone depending upon its application to a feature or to a feature-of-size. When straightness is used on a feature-of-size the following applies:

- The tolerance zone applies to the axis of the feature-of-size.

- Rule #1 is overridden.

- The virtual condition for the feature-of-size is affected.

- Modifiers may be used in the feature control frame.

- The tolerance value specified may be greater than the size tolerance.

The location of the feature control symbol on the drawing indicates whether the control applies to a feature or a feature-of-size. In Figure 2-9 the control is located so that it is directed to the pin surface. The pin surface is a feature. Therefore, the symbol applies to the surface elements of the pin. In Figure 2-10, the control is located so that it is related to the dimension of a feature-of-size. In this case, the symbol applies to a feature-of-size.

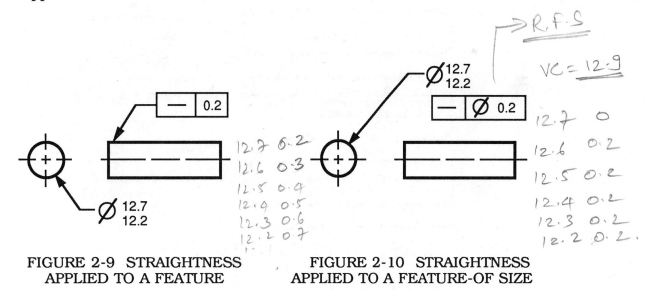

FIGURE 2-9 STRAIGHTNESS
APPLIED TO A FEATURE

FIGURE 2-10 STRAIGHTNESS
APPLIED TO A FEATURE-OF SIZE

REMEMBER

Whenever a straightness control is applied to a feature-of-size, Rule #1 is overridden and the virtual condition is affected.

43

RULE #1

Whenever Rule #1 applies to a feature-of-size an automatic straightness control exists for its axis or centerplane. This automatic control is a result of the interrelationship of Rule #1 (perfect form at MMC) and the size dimension. When the feature-of-size is at MMC, its axis or centerplane must be perfectly straight. As a feature-of-size departs from MMC, its axis or centerplane may have a straightness error, equal to the amount of departure. See Figure 2-11.

REMEMBER

Whenever Rule #1 applies to a feature-of-size, it provides an automatic straightness control for the axis or centerplane of the feature-of-size.

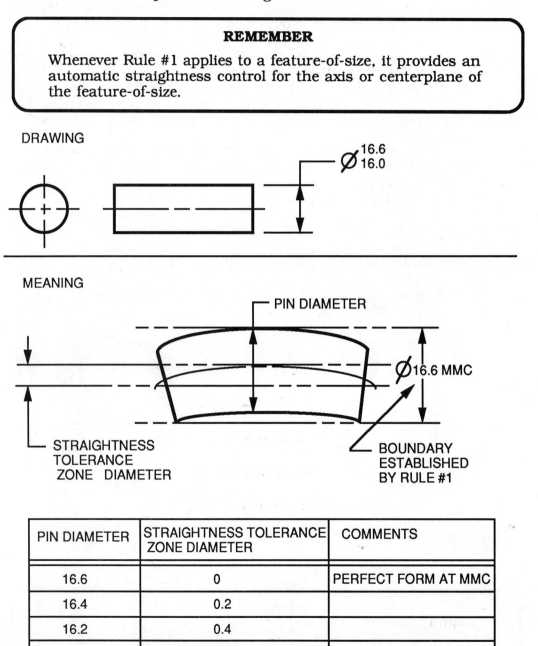

PIN DIAMETER	STRAIGHTNESS TOLERANCE ZONE DIAMETER	COMMENTS
16.6	0	PERFECT FORM AT MMC
16.4	0.2	
16.2	0.4	
16.0	0.6	

FIGURE 2-11 RULE #1 AS A STRAIGHTNESS CONTROL

APPLICATION (RFS)

When applying straightness controls to features-of-size, it must be designated if the control is to apply RFS, LMC, or MMC. This section will discuss straightness controls applied regardless of feature size. When it is desired to apply a straightness control in the RFS state, no modifier is required. This is because of Rule #3 (see Chapter 1, page 20).

When the automatic straightness control from Rule #1 is not sufficient to satisfy the functional requirements of the part, a straightness control may be added. Figure 2-12 shows an example of a straightness control applied to a feature-of-size. Note the location of the feature control symbol - it means that the control applies to a feature-of-size.

When a straightness control applies to a feature-of-size, the following applies:

- The straightness tolerance specifies a tolerance zone within which the axis or centerplane must lie.

- Rule #1 is overridden.

- Rule #3 applies.

- The virtual condition of the feature-of-size is affected.

- The feature-of-size must also be within its size tolerance.

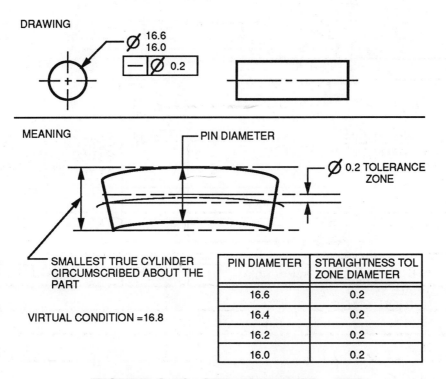

FIGURE 2-12 STRAIGHTNESS - RFS

PIN DIAMETER	STRAIGHTNESS TOL ZONE DIAMETER
16.6	0.2
16.4	0.2
16.2	0.2
16.0	0.2

VIRTUAL CONDITION =16.8

APPLICATION (MMC)

Whenever an MMC modifier is used in a straightness control, it means the stated tolerance applies when the feature-of-size is at MMC. Two important benefits become available when straightness is applied at MMC. First, as the feature size departs from MMC, a bonus tolerance becomes available. Second, whenever an MMC modifier is used in a feature control frame, the tolerance may be verified with a fixed gage. A *Fixed gage* is a gage that has no moving parts. An example is shown in Figure 2-13.

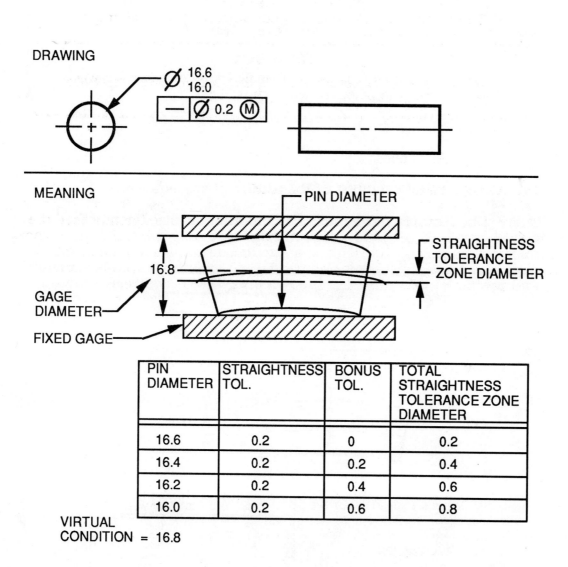

PIN DIAMETER	STRAIGHTNESS TOL.	BONUS TOL.	TOTAL STRAIGHTNESS TOLERANCE ZONE DIAMETER
16.6	0.2	0	0.2
16.4	0.2	0.2	0.4
16.2	0.2	0.4	0.6
16.0	0.2	0.6	0.8

VIRTUAL CONDITION = 16.8

FIGURE 2-13 STRAIGHTNESS - MMC

Whenever a straightness control is applied to a feature-of-size at MMC, the following applies:

- The stated straightness tolerance specifies a zone within which the axis or centerplane must lie.

- Rule #1 is overridden.

- The virtual condition of the feature-of-size is affected.

- A bonus tolerance is available.

- A fixed gage may be used to verify the part.

- The feature-of-size must also be within its size tolerance.

REMEMBER

Whenever a straightness control is applied to a feature-of-size at MMC, a bonus tolerance is available.

INDIRECT STRAIGHTNESS CONTROLS

There are a number of geometric controls that indirectly affect the straightness of an axis or centerplane of a feature-of-size. The interrelationship of Rule #1 and a size dimension is such a control. In certain cases, perpendicularity, parallelism, angularity, profile, runout, and positional tolerances are indirect straightness controls.

LEGAL SPECIFICATION TEST

For a straightness control which is applied to a feature-of-size to be a legal specification, it must satisfy the following conditions:

- No datum reference may be used in the feature control frame.

- The control must be applied to a feature-of-size.

- The straightness tolerance value specified should not be larger than the straightness tolerance resulting from any other geometric tolerances that may control the form of the feature-of-size.

A simple test for verifying if a straightness control is legal is shown in Figure 2-14.

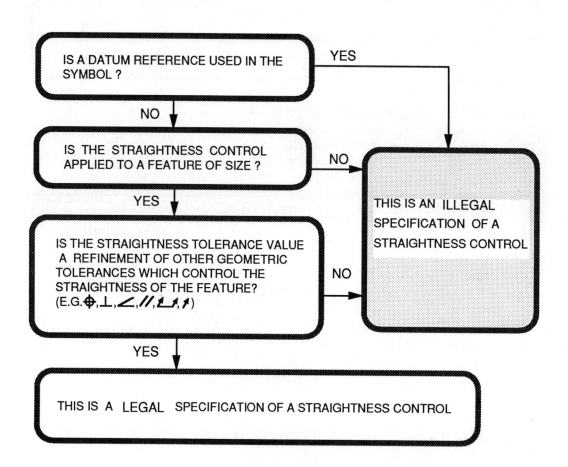

FIGURE 2-14 TEST FOR LEGAL SPECIFICATION OF STRAIGHTNESS
APPLIED TO A FEATURE-OF-SIZE

CIRCULARITY

Circularity is a condition where, at any radial section perpendicular to a common axis, the surface of a cylinder (sphere or cone) is a perfect theoretical circle. A *circularity tolerance* is the amount which surface elements of the diameter may vary from a theoretical circle. A circularity tolerance zone, applied to an external surface, consists of two concentric circles - one circumscribes the high points of the diameter and the second is radially smaller by the circularity tolerance value. See Figure 2-15. A circularity tolerance zone, applied to an internal surface, consists of two concentric circles-one contacts the low points of the diameter and the second is radially larger by the circularity tolerance value.

> **Note - The information regarding circularity controls applies to rigid parts only. For non rigid parts see ANSI Y14.5M-1982 section 6.8.**

REMEMBER

A circularity tolerance zone is two concentric circles spaced apart a radial distance equal to the circularity tolerance value.

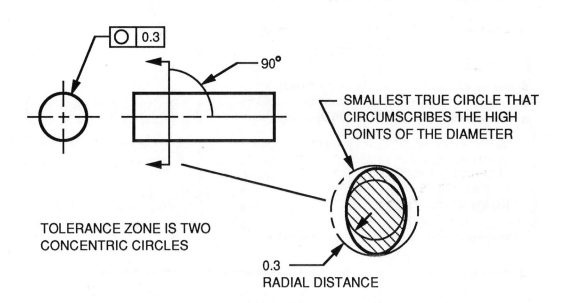

FIGURE 2-15 CIRCULARITY TOLERANCE ZONE

Circularity can only be applied to a feature (diametral surface elements), therefore, it cannot use the MMC or LMC modifier. A geometric control can only use these modifiers when it is applied to a feature-of-size. Also, Rule #3 applies.

RULE #1

Whenever Rule #1 applies to a diametral feature-of-size, an automatic circularity control exists for its surface. This automatic control is a result of the interrelationship of Rule #1 (perfect form at MMC) and the size dimension. When the diameter is at MMC, its elements must be perfectly circular. As the diameter departs from MMC, a circularity error is permissible. See Figure 2-16.

REMEMBER

Whenever Rule #1 applies to a diametral feature-of-size, it provides a circularity control for the surface elements of each diametral section.

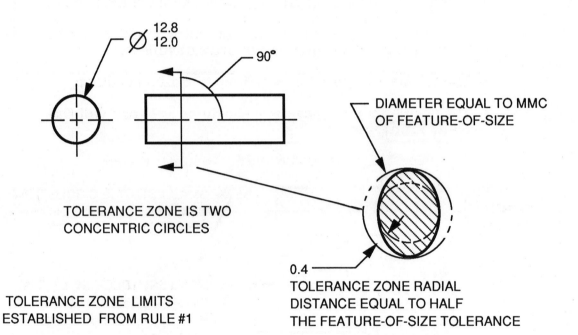

FIGURE 2-16 CIRCULARITY TOLERANCE ZONE FROM RULE #1

Figure 2-16 illustrates that whenever a diameter is controlled by Rule #1, its diametral surface elements must lie between two concentric circles, one circle equal to the MMC size of the diameter, the second circle equal to the LMC of the diameter. This is a result of the interrelationship of Rule #1 and the size dimension. Therefore, a diametral dimension automatically restricts the circularity of a diameter to one half of its size tolerance.

50

REMEMBER

Whenever Rule #1 applies to a diametral feature-of-size, the circularity is automatically restricted to one half the diametral tolerance.

APPLICATION

If the automatic indirect circularity control from Rule #1 is not sufficient to satisfy the functional requirements of the part, a circularity control may be added. A circularity control never overrides Rule #1 it refines the maximum allowable circularity error of the surface. Figure 2-17 shows an application of circularity. The following statements list concepts which apply from Figure 2-17.

- The circularity control limits the maximum surface variation only when the part departs from MMC by more than the circularity tolerance value,

- The circularity control does not override Rule #1.

- The circularity control does not affect virtual condition.

- The circularity tolerance value should be less than half the diameter tolerance.

- The diameter must also be within its size tolerance.

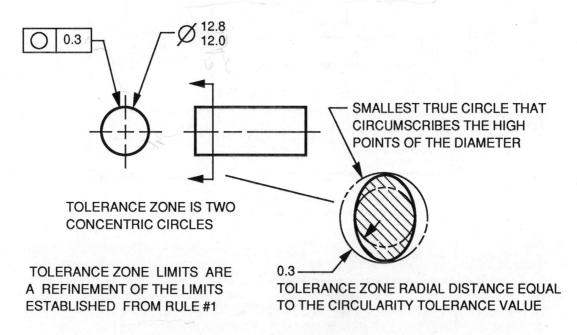

FIGURE 2-17 CIRCULARITY APPLLICATION

INDIRECT CIRCULARITY CONTROLS

There are a number of geometric controls which can affect the circularity of a surface indirectly. Rule #1, when it applies to a diameter, is such a control. In certain cases, cylindricity, profile, and runout can, also, control circularity indirectly .

LEGAL SPECIFICATION TEST

For a circularity control to be a legal specification, it must satisfy the following conditions:

- No datum reference may be used in the feature control frame.

- The control must be applied to a diametral feature.

- No modifiers may be used.

- The circularity tolerance value specified must be a refinement of any other geometric tolerances that control the circularity of the feature (e.g. Rule #1, cylindricity, profile, and runout.)

A simple test for verifying if a circularity control is specified correctly is shown in Figure 2-18.

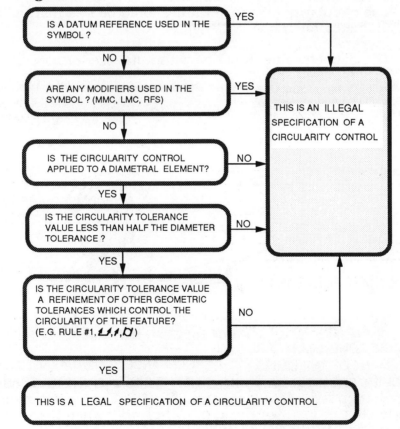

FIGURE 2-18 TEST FOR LEGAL SPECIFICATION OF CIRCULARITY

CYLINDRICITY

A *cylindricity tolerance* is the amount which surface elements of a cylinder may be allowed to vary from a theoretically perfect cylinder. A cylindricity tolerance zone consists of two concentric cylinders. Applied to an external cylindrical feature, one tolerance zone cylinder circumscribes the high points of the diameter and the second is radially smaller by the cylindricity tolerance value. See Figure 2-19. Applied to an internal cylindrical feature, one tolerance zone cylinder contacts the high points of the diameter, and the second is radially larger by the cylindricity tolerance value.

REMEMBER

A cylindricity tolerance zone is two concentric cylinders spaced apart a radial distance equal to the cylindricity tolerance value.

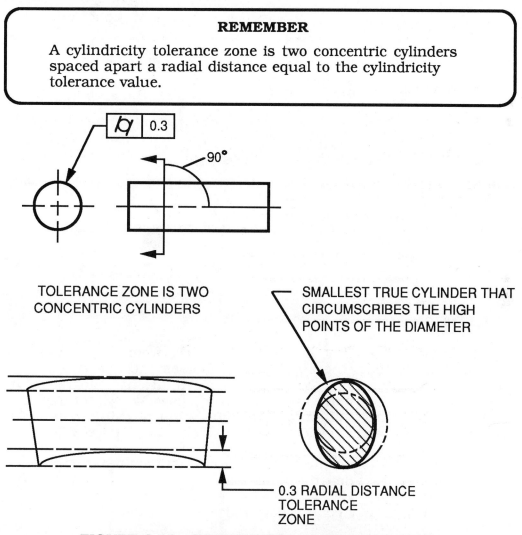

TOLERANCE ZONE IS TWO
CONCENTRIC CYLINDERS

SMALLEST TRUE CYLINDER THAT
CIRCUMSCRIBES THE HIGH
POINTS OF THE DIAMETER

0.3 RADIAL DISTANCE
TOLERANCE
ZONE

FIGURE 2-19 CYLINDRICITY TOLERANCE ZONE

Cylindricity can only be applied to a feature (surface elements of a cylinder), therefore, it cannot use the MMC or LMC modifier. A geometric control can use these modifiers only when it is applied to a feature-of-size. Also, Rule #3 applies.

RULE #1

Whenever Rule #1 applies to a cylindrical feature-of-size, an automatic cylindricity control exists for its surface. This automatic control is a result of the interrelationship of Rule #1 (perfect form at MMC) and the size dimension. When the cylinder is at MMC, it must be a perfect cylinder (perfect circularity, perfect straightness, and perfect cylindricity).

> **REMEMBER**
>
> Whenever Rule #1 applies to a cylindrical feature, it provides a cylindricity control for the surface elements of the cylinder.

Whenever a cylindrical feature-of-size is controlled by Rule #1, its surface elements must lie between two concentric cylinders, one cylinder equal to the MMC of the feature-of-size, the second cylinder equal to the LMC of the feature-of-size. This is a result of the interrelationship of Rule #1 and the size dimension. Therefore, a diametral dimension automatically restricts the cylindricity of a cylindrical feature-of-size to half of the size tolerance.

> **REMEMBER**
>
> For cylindrical features-of-size, the cylindricity is automatically restricted to one half of the diametral tolerance.

APPLICATION

If the automatic indirect cylindricity control from Rule #1 is not sufficient to satisfy the functional requirements of the part a cylindricity control may be added. A cylindricity control never overrides Rule #1 - it refines the maximum allowable cylindricity error of the part. Figure 2-20 shows an application of cylindricity. The following statements list concepts which apply from Figure 2-20.

- The cylindricity control limits the maximum surface variation when the part departs from MMC.

- The cylindricity control does not override Rule #1.

- The cylindricity control does not affect virtual condition.

- The cylindricity tolerance value should always be less than half the diameter tolerance.

- The cylindricity control is a composite control which includes circularity, straightness and taper of a cylindrical feature.

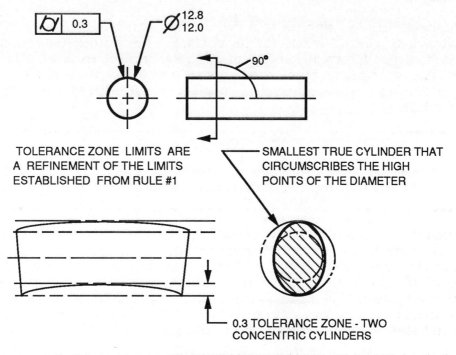

TOLERANCE ZONE LIMITS ARE
A REFINEMENT OF THE LIMITS
ESTABLISHED FROM RULE #1

SMALLEST TRUE CYLINDER THAT
CIRCUMSCRIBES THE HIGH
POINTS OF THE DIAMETER

0.3 TOLERANCE ZONE - TWO
CONCENTRIC CYLINDERS

FIGURE 2-20   CYLINDRICITY APPLICATION

INDIRECT CYLINDRICITY CONTROLS

There are a number of geometric controls which can affect the
cylindricity indirectly. When applicable, the interrelationship of Rule
#1 and a diameter dimension , is such a control. In certain cases, total
runout, positional tolerance, and profile can also control cylindricity
indirectly.

LEGAL SPECIFICATION TEST

For a cylindricity control to be a legal specification, it must satisfy the
following conditions:

- No datum references may be used in the feature control frame.

- The control must be applied to a cylinder.

- No modifiers may be used.

- The tolerance value specified must be a refinement of any
  other geometric tolerance that controls the cylindricity of the
  feature.

A simple test for verifying if a cylindricity control is legal is shown in
Figure 2-21.

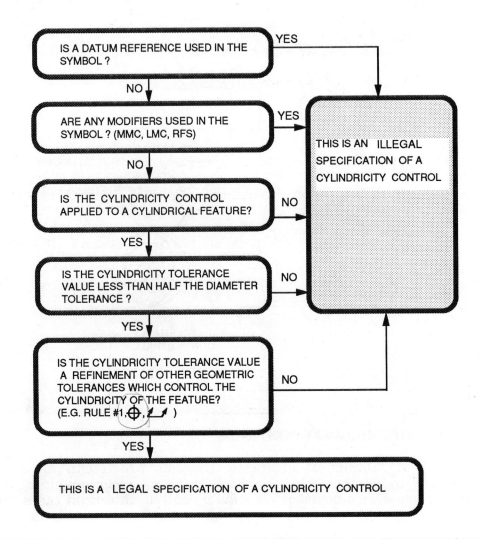

FIGURE 2-21 TEST FOR LEGAL SPECIFICATION OF CYLINDRICITY

SUMMARY

A summary of form control information is shown in Figure 2-22.

	DATUM REFERENCE REQ'D OR PROPER	CAN BE APPLIED TO		AFFECTS VIRTUAL CONDITION	CAN USE MMC MODIFIER	CAN OVERRIDE RULE #1
		FEATURE	FEATURE OF SIZE			
FLATNESS ▱	NO	YES	NO	NO	NO	NO
STRAIGHTNESS —	NO	YES	YES	YES*	YES*	YES*
CIRCULARITY ○	NO	YES	NO	NO	NO	NO
CYLINDRICITY ⌭	NO	YES	NO	NO	NO	NO

* ONLY WHEN APPLIED TO A FEATURE-OF-SIZE

FIGURE 2-22 SUMMARIZATION OF FORM CONTROLS

VOCABULARY WORDS

Flatness Tolerance
Straightness Tolerance (of a feature)
Fixed Gage
Straightness Tolerance (of a feature-of-size)
Circularity Tolerance
Cylindricity Tolerance

THOUGHT QUESTIONS

1. Should the inspection department inspect the form of every feature-of-size on a part?

2. Why is it improper to use a datum reference with a form control?

PROBLEMS AND QUESTIONS

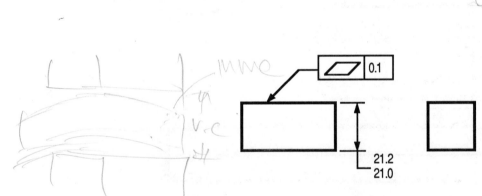

QUESTIONS 1-7 REFER TO THE FIGURE ABOVE

1. Each point of the top of the block must lie between two parallel planes spaced ____0.1____ apart.

2. The reference plane for measuring the flatness of the top of the block is established by the three ___high___ points of the surface.

3. If the size tolerance is increased, will it change the flatness tolerance zone? ___yes___

4. Is it permissible to use an MMC modifier with the flatness symbol? _____NO_____

5. The flatness tolerance must be _____less_____ than the size tolerance.

6. The virtual condition of the height of the block is _____21.3_____.

7. If the block was at MMC, what would be the allowable flatness error on the top surface? _____0_____

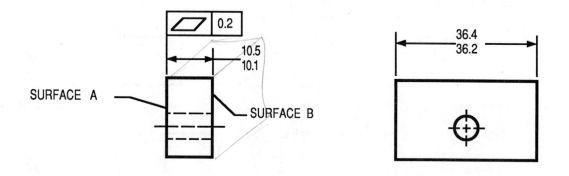

QUESTIONS 8-14 REFER TO THE FIGURE ABOVE

8. What is the virtual condition of the 10.1-10.5 dimension?
_____10.7_____

9. What is the virtual condition of the 36.2 - 36.4 dimension?
_____36.4_____

10. If the part was at MMC, what would be the allowable flatness error of surface A? _____

11. Each point of surface A must lie between two parallel planes _____0.2_____ apart.

12. Does Rule #1, or the flatness tolerance, limit the permissible flatness error of surface A when the part is at LMC?

_____

13 Does Rule #1, or the flatness tolerance, limit the permissible flatness error of surface A when the part is at MMC?

_____

14. What is the maximum flatness error permissible on surface B?

_____

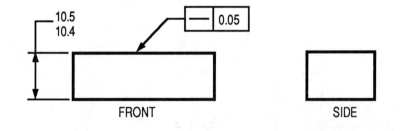

QUESTIONS 15-19   REFER TO THE FIGURE ABOVE

15. What is the virtual condition of the height of the block?
    ___10.5___

16. What limits the straightness of the line elements in the side view? ___0.05___

17. If the part was at MMC, what would be allowable straightness error for the line elements in the front view? ___0.05___

18. If the straightness symbol was revised to ⎯ 0.05 Ⓜ , what effect would this have on the part? ___Wrong.___

19. If the straightness symbol was revised to ⎯ 0.15 , what effect would this have on the part? ___RFS the st tolerance___
    ___is 0.15___

QUESTIONS 20-23   REFER TO THE FIGURE ABOVE

20. The virtual condition of the pin diameter is __12.54__.

21. If the pin was at MMC, what would be the allowable straightness error of the line elements? __0__

22. Fill in the chart

ACTUAL PIN DIAMETER	ALLOWABLE STRAIGHTNESS TOL.
11.9	
12.0	
12.2	
12.4	
12.5	
12.6	

23. If the pin was at LMC, what would be the allowable straightness error of the line elements? __0.04__

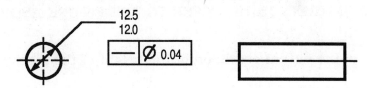

QUESTIONS 24-26   REFER TO THE FIGURE ABOVE

24. The virtual condition of the pin diameter is __12.54__.

25. Does the straightness control apply to the surface elements of the pin? __NO__ Why? __It is a part of a size control frame and applies to axis__

26. Fill in the chart. (For each condition in the chart fill in the allowable straightness error.)

ACTUAL PIN DIAMETER	USING THE SYMBOL SPECIFIED	ADDING AN MMC MODIFIER TO THE SYMBOL	REMOVING THE SYMBOL
12.5	0.04	0.08	0
12.46	0.04	0.10	0.04
12.42	0.04	0.12	0.08
12.40	0.04	0.14	0.1
12.36	0.04	0.16	0.14

60

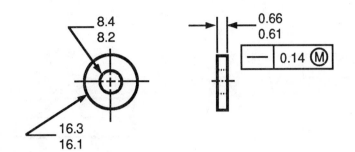

QUESTIONS 27-29 REFER TO THE FIGURE ABOVE

27. What is the virtual condition of the thickness of the
    washer?_____

28. If the washer is at LMC, what is the allowable straightness error?

    _____

29. Does Rule #1 apply to the thickness of the washer?

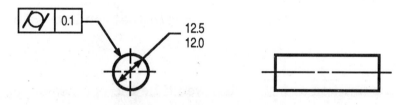

QUESTIONS 30-33 REFER TO THE FIGURE ABOVE

30. Describe the tolerance zone for the cylindricity callout. _____

    _____

    _____

31. The virtual condition of the pin is _____ .

32. What is the maximum circularity error permissible on the pin?

    _____

33. If the pin was at MMC, what would be the permissible cylindricity
    error? _____

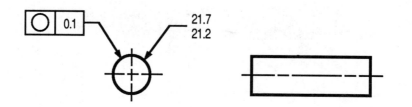

QUESTIONS 34-38 REFER TO THE FIGURE ABOVE

34. Describe the tolerance zone for the circularity callout. _____
    _____

35. What is the largest value the circularity control could be and still
    make sense? _____

36. If the circularity control was left off, what would control the
    circularity limits? _____

37. Can a circularity control affect the virtual condition of a
    diameter? _____

38. If the pin was at MMC what would be the permissible circularity
    error? _____

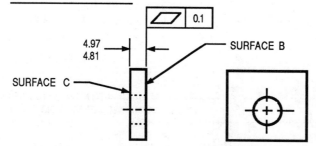

39. Using the drawing above, fill the chart below

IF THE SYMBOL AT FLATNESS WAS	MAX ALLOWABLE FLATNESS ERROR ON SURFACE ...		MIN SETTING OF PARALLEL PLATES WHICH THE PART WILL PASS THRU	VIRTUAL CONDITION OF PLATE THICKNESS
	B	C		
SAME AS SPECIFIED				
REVISED TO READ ▱ 0.1 C				
REMOVED				

ACROSS

3. STUN
5. RODENT
8. FORM CONTROLS NEVER USE ONE
9. EMBROIDER
10. LARGE BUNDLE
12. LESS THAN WHOLE
13. PERFORM
16. STILL
17. PERFECT FORM AT MMC
18. EXTRATERRESTRIAL
20. STRAIGHTNESS CAN CONTROL AN _____
21. FLATNESS TOLERANCE ZONE
22. TWO _____ PLANES
23. OBESE
26. CONTROL WHICH AFFECTS VIRTUAL CONDITION
28. TRAVEL
29. ONE OF A PAIR
30. CALL _____ DAY
34. TYPE OF TOLERANCE
36. LINE _____
37. SHARPEN
41. SLANT
43. TECHNICAL BOOK AUTHOR
45. EFFECT OF FLATNESS ON A SIZE TOLERANCE
46. HALF
47. DINE
48. KIND OF PAPER
49. NUMBER OF FORM CONTROLS
50. CYLINDRICITY APPLIES TO THESE ELEMENTS OF A DIAMETER

DOWN

1. SAY
2. CASTLE DEFENSE
3. FARM BUILDING
4. SYMBOL FOR CIRCULARITY
6. MAO _____-TUNG
7. PRIZE
8. CIRCULARITY APPLIES TO _____ PART FEATURES
11. DISTRICT
14. FORM CONTROLS
15. A GAGE WITH NO MOVING PARTS
19. LOCATES FLATNESS TOLERANCE ZONE
22. HOLE
24. EASTER _____
25. EYES
26. SNAKE
27. 60 MINUTES
31. FLATNESS ALWAYS APPLIES IN THIS CONDITION (ABBR)
32. STRAIGHTNESS AT MMC ALLOWS THIS TOLERANCE
33. PLANAR SURFACE CONTROL
35. FLATNESS CAN NEVER USE A _____
38. INDIRECT FORM CONTROL
39. REGULATION
40. STRAIGHTNESS OF A _____
42. ENAMEL
43. RULER
44. 12 MONTHS

chü rg

CHAPTER **3**

DATUMS

Say what you mean.

Rudolf Flesch

INTRODUCTION

This chapter introduces the terminology and concepts which relate dimensional measurements to theoretical planes (or axes) called datums. This information is essential in order to understand how dimensions and their tolerances should be measured.

GENERAL INFORMATION

What is a Datum?

A *datum* is a theoretically exact point, line, axis or plane which indicates the origin of a specified dimensional relationship between a toleranced feature and a designated feature on a part. A designated part feature serves as the datum feature, whereas, its true counterpart (the gage) establishes the datum plane or axis. For practical reasons, a datum is assumed to exist and be simulated by processing or inspection equipment, such as machine tables, surface plates, collets, gage surfaces, etc.

[handwritten margin note: Indicates the origin of a specified dimensional relationship btwn a toleranced feature & a designated feature on a part]

What is the Purpose of a Datum?

[handwritten note: X To locate a feature of a part for checking geometric tolerance related to the datum feature]

Datums are used primarily to locate a part in a repeatable manner for checking geometric tolerances related to the datum features.

In addition, datums communicate functional design information about the part. For example, the datum features on a drawing show the drawing users, which part features mount and locate the part in its assembly. Also, the primary datum feature is the part feature which establishes the attitude of the part in its assembly.

What is a Datum Feature?

A datum feature is a part feature which contacts, or is used to establish, a datum.

How are Datums Specified and Referenced?

The symbol, for specifying a datum feature, is a rectangle containing the datum identifying letter preceded and followed by a dash. See Figure 3-1.

Datums are referenced in feature control frames. In a feature control frame, the compartment next to the tolerance value compartment references a primary datum. Secondary and tertiary datum references follow.* See Figure 3-2.

REMEMBER

Datum features are part features and datums are theoretical reference planes or axes.

66

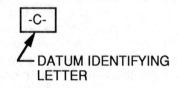

DATUM IDENTIFYING
LETTER

FIGURE 3-1 SPECIFYING A DATUM

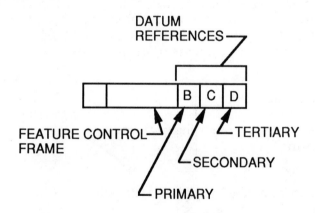

FIGURE 3-2 REFERENCING DATUMS

How are Datum Features Selected?

Datum features are selected on the basis of the functional design requirements of a part. The datum features are the surfaces which locate and mount the part in its assembly. For example, the part shown in figure 3-3, mounts on surface A and is located by surface B. For assembly, the bolt holes need to be held relative to the features which locate and mount the part. Therefore, surface A and diameter B are designated as datum features and the bolt holes are dimensioned relative to datums A and B thru the use of geometric tolerances. Since the part is clamped against surface A, this surface will establish the attitude of the part and is referenced as the primary datum in the dimensioning of the bolt holes. The pilot diameter locates the part and is referenced as the secondary datum, in the dimensioning of the bolt holes.

> **REMEMBER**
>
> Datums are selected on the basis of functional design requirements of the part.

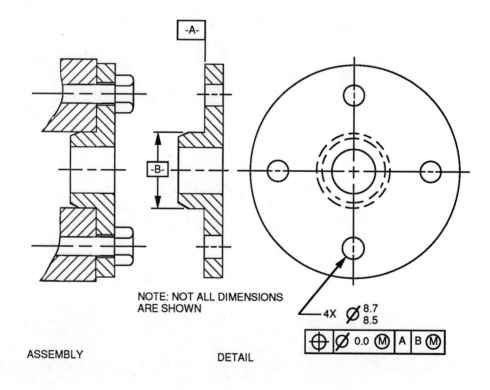

FIGURE 3-3 DATUM SELECTION EXAMPLE

PLANAR FEATURE DATUMS

This section deals with planar feature datums only. In some cases, for the measurements required, a single datum reference is sufficient. When this is the case, as shown in Figure 3-4, the datum reference is considered a primary datum. A primary datum always establishes the attitude of the part for measurement. A *datum plane* is a theoretical plane which contacts the three high points of the datum feature. Measurements from a datum plane are always made perpendicular to the datum plane. When measuring geometric tolerances which reference the datum, the three high points of the datum feature (surface) must be contacting the datum plane. Only part dimensions which are related to a datum thru geometric tolerances or special notes* should be measured from the datum plane. See Figure 3-4.

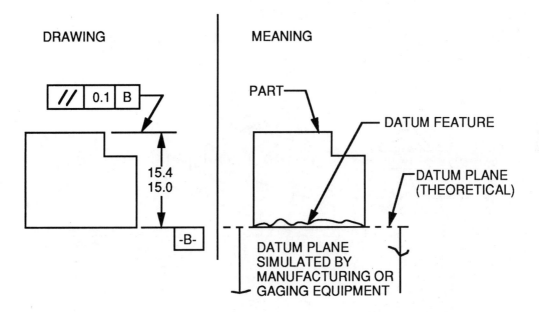

FIGURE 3-4 SURFACE AS A PRIMARY DATUM

* Rule #1 is such a note. Under Rule #1, any feature-of-size acts as an implied datum (relative to itself). The implied self datum is used to measure the size of the feature-of-size. Caution- Rule #1 does not relate a dimension to a datum reference frame (because it does not specify a datum precedence.) See page 70.

DATUM REFERENCE FRAME

When more than one datum plane is required for repeatable measurements, a datum reference frame is used. A *datum reference frame* is a set of three mutually perpendicular planes. See Figure 3-5. These planes provide direction as well as an origin for measurements. For specified measurements, the part datum features contact the datum planes.

The planes of a datum reference frame are exactly 90° to each other by definition, but, the actual surfaces of the part must have an angular tolerance specified on the drawing. See Figure 3-6. When making part measurements that are related to a datum reference frame, the part must be brought into contact with the datum reference frame in a prescribed manner. The part feature contacting the datum reference frame first is the *primary datum.** The part feature contacting the datum reference frame second is the *secondary datum.** The part feature contacting the datum reference frame third is the *tertiary datum.** Feature control symbols specify which datum feature is primary, secondary, and tertiary.

REMEMBER

The interrelationship between datum features must be specified.

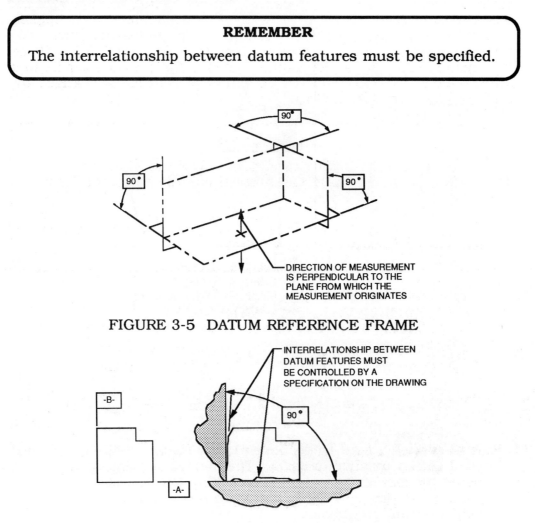

FIGURE 3-5 DATUM REFERENCE FRAME

FIGURE 3-6 RELATIONSHIP OF DATUM FEATURES

70

DATUM PRECEDENCE

In order to position a part on a datum reference frame, in a repeatable manner, datums must be referenced in an order of precedence. *Datum precedence* refers to the order in which part features come in contact with the datum reference frame (1st, 2nd, 3rd). Figure 3-7 is an example of a part where the datum features are planar. The intended datum precedence is indicated by the order of the datum reference letters in the feature control frame. The datum feature surfaces are identified as D, E, and F. (These surfaces represent the functional locating surfaces of the part.) When checking the location of the holes, datum D is primary, datum E is secondary, and datum F is tertiary. Only dimensions which are related to the datum reference frame thru geometric tolerances or special notes should be measured from the datum reference frame.

A primary datum feature is related to the datum reference frame by bringing a minimum of three points of the datum feature surface in contact with the primary datum plane. See Figure 3-8A. The part is further related to the datum reference frame by bringing at least two points of the secondary datum feature (surface E) into contact with the second datum plane. See Figure 3-8B. The relationship is completed by bringing at least one point of the tertiary datum feature (surface F) into contact with the third datum plane. See Figure 3-8C. These relationships establish a repeatable method of addressing a part to a datum reference frame. (Note, if the datum precedence is changed the hole locations would be different.)

> **REMEMBER**
>
> Only dimensions related to the datum reference frame thru geometric tolerances or special notes should be measured from the datum reference frame.
>
> Every dimension that is related to a datum reference frame must have a datum precedence specified.

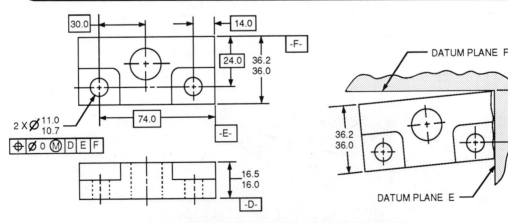

FIGURE 3-7 DATUM PRECEDENCE

3-2-1 RULE

The *3-2-1 Rule* defines the minimum number of points of contact required for a primary, secondary, and tertiary datum feature with their respective datum planes. The primary datum feature requires at least three points of contact with its datum plane. The secondary datum feature requires at least two points of contact with its datum plane. The tertiary datum feature requires at least one point of contact with its datum plane. The 3-2-1 Rule applies to planar datum features only.

REMEMBER

The 3-2-1 Rule defines the minimum number of points of contact for the primary datum as 3, the secondary datum as 2, and tertiary datum as 1.

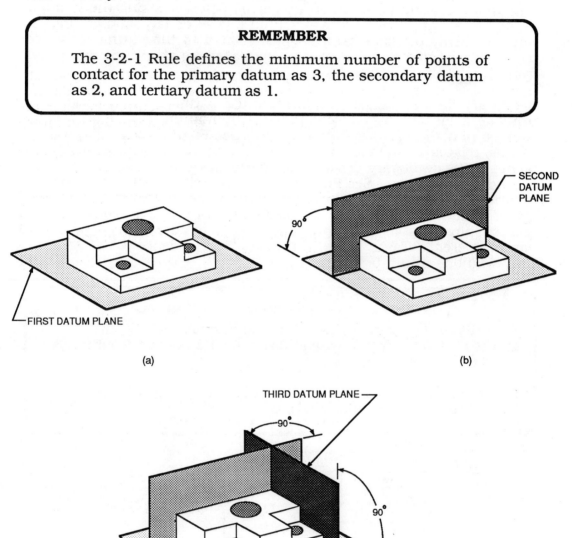

(a)

(b)

(c)

FIGURE 3-8   3-2-1 RULE

DATUM TARGETS

Datum targets are designated points, lines or areas of contact used to locate a part in a datum reference frame.* Datum targets are shown on the part surfaces on a product drawing, but they describe the shape and location of the gage features that are used to simulate the datum planes.

Datum targets should be considered for use whenever the whole surface may introduce uncertainties of obtaining repeatable measurements. Typical places where datum targets are desirable are: castings, forgings, warped or bowed surfaces which may rock when in contact with a theoretically flat plane.

Datum target points, lines, and areas are identified by using a datum target symbol. See Figure 3-9. A solid leader line, from the datum target symbol to the part surface, indicates that the datum target is on the near (visible) surface. A dashed leader line, from the datum target symbol to the part surface, indicates that the datum target is on the far (hidden) surface. In addition, the datum feature should be identified with a datum designation symbol. See Figure 3-10.

REMEMBER

Datum targets describe the shape, size, and location of *gage* features which are used to establish planes.

FIGURE 3-9 DATUM TARGET SYMBOL

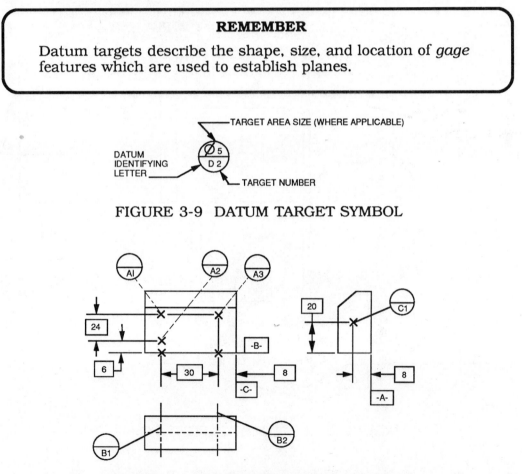

FIGURE 3-10 DATUM TARGET EXAMPLE

APPLICATIONS

A datum target point is specified by an " X " symbol. The symbol is shown and dimensioned on a view which shows the plan view of the surface on which it is being applied. Where this type of a view is not available, the symbol can be shown and dimensioned in two adjacent views. Basic dimensions should be used to locate the datum target points relative to each other and the other datums on the part. See Figure 3-11. The use of basic dimensions requires each gage to be built to exactly* the same dimensions. Identical gages are necessary for consistent results when inspecting parts.

REMEMBER

Basic dimensions should be used to describe the shape, size, and location of datum targets.

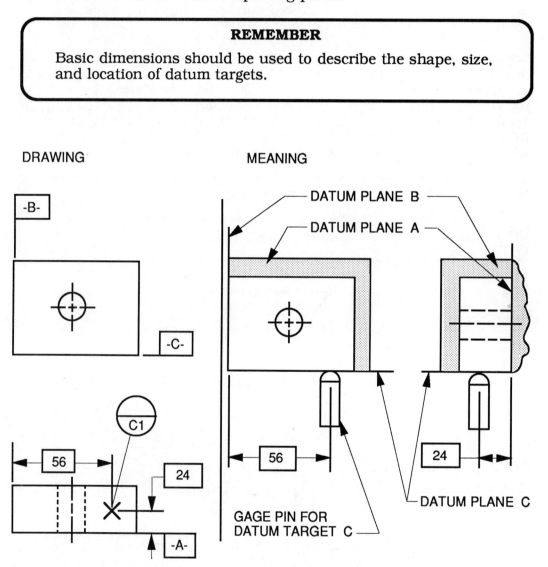

FIGURE 3-11   DATUM TARGET POINT

* See discussion on basic dimensions in Chapter One.

When it is desired to use a line contact between the datum feature and the datum plane, a datum target line is specified. A datum target line can be specified three ways; a phantom line on the plan view of the surface (See Figure 3-12A), an " X " symbol on the edge view of the surface (See Figure 3-12B), and a combination of both of the above (See Figure 3-12C). In each method, basic dimensions should be used to locate the datum target relative to other targets and/or datums. Figure 3-13 illustrates the application of a datum target line.

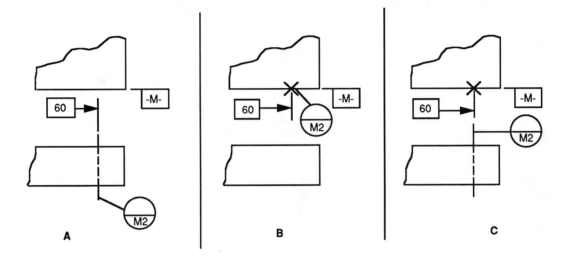

FIGURE 3-12   DATUM TARGET LINES

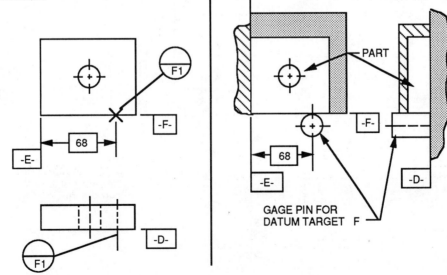

FIGURE 3-13   DATUM TARGET LINE

When it is desired to use a specific area or areas to contact a datum plane, a target area or areas of the desired shape and size is specified. A datum target area is designated by section lines within a phantom outline of the target area, with the necessary basic dimensions added to describe shape and location. See Figure 3-14A. If the target area is circular, the diameter may be specified in the upper half of the datum target symbol. See Figure 3-14B. When it is impractical to show the circular target area, the method shown in Figure 3-14C may be used. Figure 3-15 illustrates an application of datum target areas.

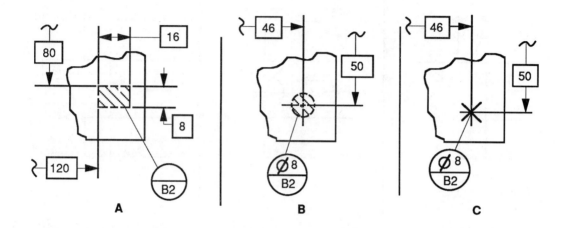

FIGURE 3-14   DATUM TARGET AREAS

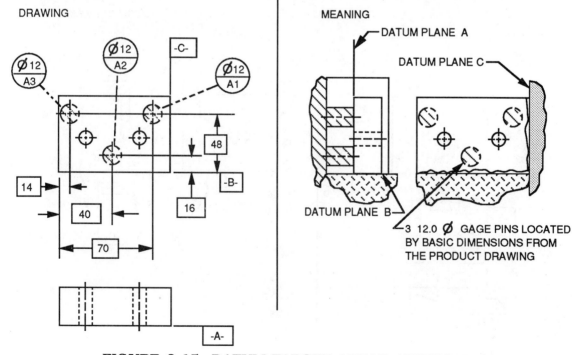

FIGURE 3-15   DATUM TARGET AREAS APPLIICATION

FEATURE-OF-SIZE DATUMS

When referencing a feature-of-size datum, it is necessary to specify which way the datum feature is to be simulated, that is: LMC, RFS, or MMC. This is accomplished thru the use of modifiers that appear in a feature control frame which references the datum feature. Note a datum feature may be referenced in several conditions in different feature control frames on the same drawing. Rule #3 specifies an automatic modifier in some cases (see Chapter 1 page 20).

When a planar surface is specified as a datum feature, it is used to establish a datum plane. Where a feature-of-size is specified as a datum feature, the surface or surfaces of that feature-of-size are used to establish a datum axis or centerplane.

REMEMBER

Whenever a datum feature-of-size is referenced it must be specified if it applies at LMC, RFS, or MMC.

APPLICATIONS (RFS)

When a feature-of-size is referenced as a datum feature on an RFS basis, the datum axis or centerplane is established thru physical contact between the datum feature surface(s) and the inspection equipment. Devices which are variable in size, such as a chuck, collet, or centering device, are used to simulate the true geometric counterpart of the datum feature and establish the datum axis or centerplane.

The following paragraphs describe the four most common applications of datum features-of-size referenced RFS.

1. Diameter as a primary datum feature (RFS)

    For an external diameter, the datum axis is the axis of the smallest true cylinder which contacts the high points of the diameter's surface. See Figure 3-16.

    For an internal diameter, the datum axis is the axis of the largest true cylinder which contacts the high points of the diameter's surface. See Figure 3-17.

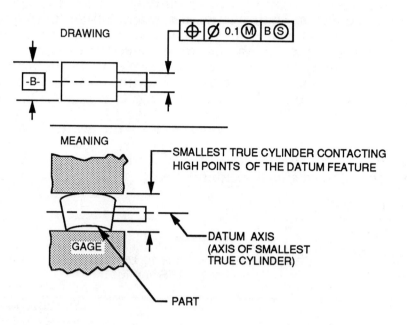

FIGURE 3-16   EXTERNAL DIAMETER AS A PRIMARY DATUM (RFS)

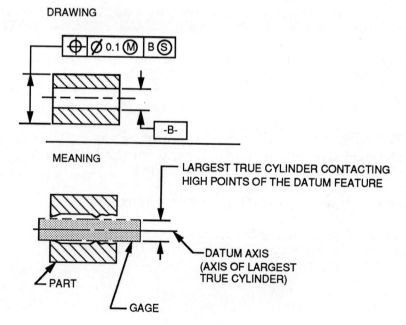

FIGURE 3-17   INTERNAL DIAMETER AS A PRIMARY DATUM (RFS)

2.   Planar feature-of-size as a primary datum feature (RFS)

For an internal planar feature-of-size, the datum plane is the centerplane of two parallel planes which contact the high points of the feature-of-size surfaces. See Figure 3-18. For an external planar feature-of-size, the datum plane is the centerplane of two parallel planes which contact the high points of the feature-of-size surfaces. See Figure 3-19. Note, these applications should be used only when the datum feature is long enough to establish the attitude of the part.

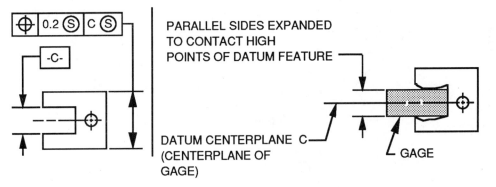

FIGURE 3-18
INTERNAL PLANAR FEATURE-OF-SIZE AS PRIMARY DATUM (RFS)

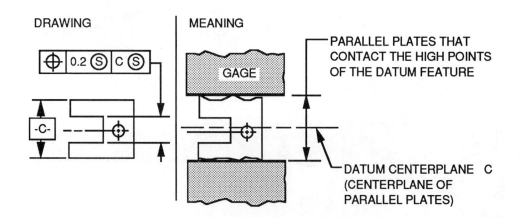

FIGURE 3-19
EXTERNAL PLANAR FEATURE-OF-SIZE AS PRIMARY DATUM (RFS)

3. Diameter or planar feature-of-size as a secondary datum feature (RFS)

    Whenever an external or internal feature-of-size is referenced as a secondary datum, the axis or centerplane is established in the same manner as indicated in no. 1 or 2 above with one additional requirement - the contacting true cylinder (or parallel planes) are, by definition, perpendicular to the primary datum plane (or axis). Figure 3-20 illustrates this principle using a diameter.

DRAWING

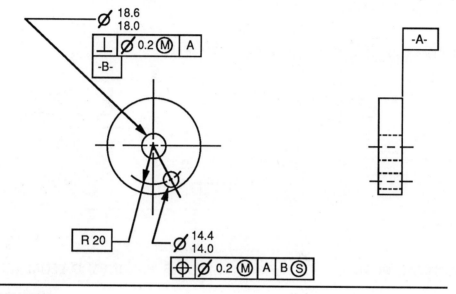

MEANING

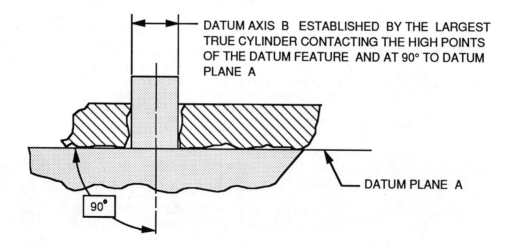

DATUM AXIS B ESTABLISHED BY THE LARGEST TRUE CYLINDER CONTACTING THE HIGH POINTS OF THE DATUM FEATURE AND AT 90° TO DATUM PLANE A

DATUM PLANE A

FIGURE 3-20   DIAMETER AS A SECONDARY DATUM (RFS)

80

4. Diameter or planar feature-of-size as a tertiary datum feature (RFS)

Whenever an external or internal feature-of-size is referenced as a tertiary datum, the axis or centerplane is established in the same manner as indicated in no. 1 or 2 above with an additional requirement - the contacting true cylinder (or parallel planes) must be oriented in relation to both the primary and secondary datum planes (or axes). The tertiary datum feature may be aligned with, or offset from, a plane of the datum reference frame. See datum C in Figure 3-21. This figure illustrates the principle using a planar feature-of-size.

DRAWING

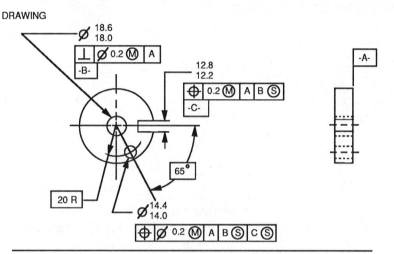

MEANING

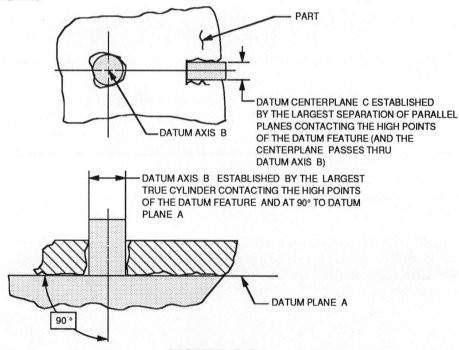

FIGURE 3-21
PLANAR FEATURE OF SIZE AS A TERTIARY DATUM (RFS)

APPLICATIONS (MMC)

Where a datum feature-of-size is referenced at MMC, the gaging equipment which is used to simulate the true geometric counterpart and establish the datum, is a fixed size. The datum axis (or centerplane) is the axis (or centerplane) of the gaging equipment and not the part axis. See Figure 3-22. The size of the gage for simulating an MMC datum is determined by the MMC size of the datum feature.

There are several instances where a datum feature-of-size is referenced at MMC and the virtual condition is used to determine the gage size. These are called *Virtual Condition Datums*. Two examples of Virtual Condition Datums are:

Example 1

Where a primary datum feature-of-size is referenced in a geometric frame at MMC, but has a geometric tolerance applied to it, the MMC rule is overruled and the datum feature will be simulated at its virtual condition.

Example 2

Where a datum feature-of-size is referenced at MMC in a feature control frame as secondary or tertiary and has other feature control symbols relating it back to the same primary datum, the MMC rule is overruled and the datum feature will be simulated at its virtual condition.

Figures 3-23 & 3-24 show examples 1, and 2 respectively.

REMEMBER

Whenever a datum feature-of-size is referenced at MMC, it applies at its virtual condition if either of the two conditions described above exist.

It may be helpful to compare a datum feature-of-size referenced at RFS and at MMC. In an RFS application, the gaging equipment is variable in size (such as a movable chuck), and physical contact is made between the datum feature surface(s) and the gaging equipment. The part is held securely in the gage - no part movement is allowed. In an MMC application, the gaging equipment is fixed in size and the part may not necessarily be held securely in the gage.

DRAWING

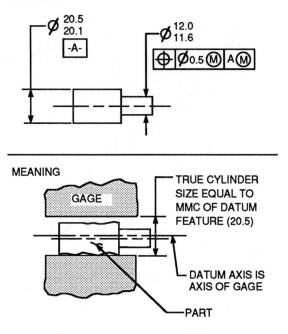

MEANING

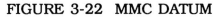

FIGURE 3-22   MMC DATUM

DRAWING

MEANING

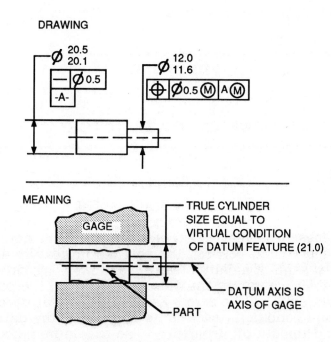

FIGURE 3-23   MMC DATUM SIMULATED AT VIRTUAL CONDITION

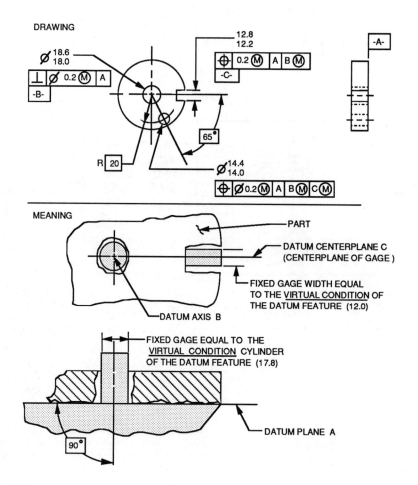

FIGURE 3-24   DIAMETER AS A SECONDARY DATUM (MMC)

DATUM SHIFT

When gaging a part with a datum feature-of-size referenced at MMC, the gage is a fixed size. Since the size of the actual part datum feature can vary, a looseness may exist between the part and gage.   This looseness or movement, between the part datum feature and the gage, is called *datum shift*. If a datum feature is at MMC (or virtual condition where applicable), no movement will occur between the part and gage, and the datum shift will be zero. As the datum feature departs from MMC (or virtual condition) towards its LMC limit, the datum shift will be equal to the amount of departure. The maximum amount of datum shift is equal to the datum feature MMC (or virtual condition), minus the datum feature LMC. See Figure 3-25.

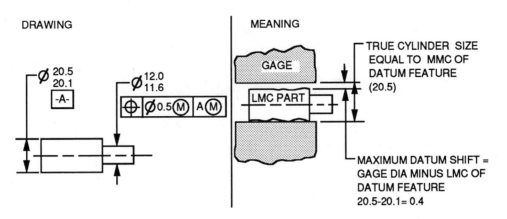

DRAWING

MEANING

Ø 20.5 / 20.1
-A-

Ø 12.0 / 11.6

⊕ Ø0.5Ⓜ AⓂ

GAGE

LMC PART

TRUE CYLINDER SIZE EQUAL TO MMC OF DATUM FEATURE (20.5)

MAXIMUM DATUM SHIFT = GAGE DIA MINUS LMC OF DATUM FEATURE
20.5-20.1= 0.4

FIGURE 3-25   DATUM SHIFT

REMEMBER

The maximum amount of datum shift is the difference between the gage size and the LMC of the datum feature.

Functional requirements of a part determine when a datum should be referenced at RFS or MMC. When a datum feature is referenced at MMC, the inspection equipment is usually simpler and less expensive. Also, since datum shift is permitted, the part may be more economical to manufacture.

CO-DATUMS

When two datum features of equal importance are used to establish a single datum plane or axis, they are called *co-datums*. A co-datum is designated by placing the appropriate datum reference letters, separated by a dash, into the datum reference compartment of the feature control frame. See Figure 3-26.

In Figure 3-27, datum plane A-B is established by a theoretical plane contacting the high points of both datum feature surfaces A and B simultaneously.

In Figure 3-28, datum axis A-B is established by simultaneously contacting the high points of both diameters with two co-axial true cylinders.

NOTE - Co-datums are used with co-axial and co-planar features only. Co-datums are normally used in primary datum applications.

REMEMBER

Co-datums are used when both datum features have an equal role in locating the part in its assembly.

85

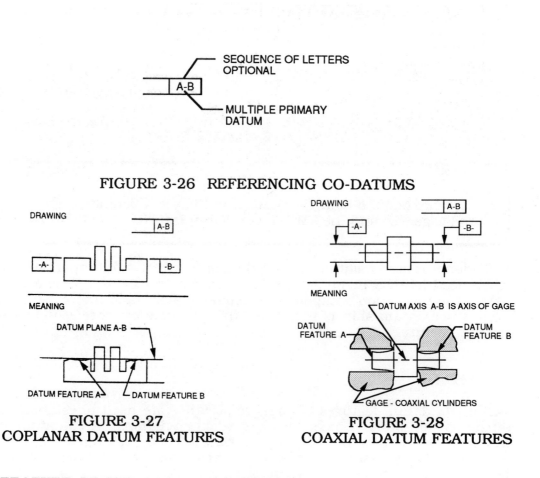

FIGURE 3-26 REFERENCING CO-DATUMS

FIGURE 3-27
COPLANAR DATUM FEATURES

FIGURE 3-28
COAXIAL DATUM FEATURES

FEATURE-OF-SIZE DATUM PRECEDENCE

When interpreting datum reference frames that involve both feature-of-size datums and planar datums the datum precedence plays a major role in the final part tolerances.

In the top portion of Figure 3-29, the datum portion of the feature control frame is left blank. In the bottom portion of the Figure the datum reference frame is completed in three different methods.

In Panel A the datum precedence is A primary RFS and B secondary.

- A moveable gage is required and no datum shift is permissible on datum feature A.

- Datum feature B will have one point contact with its datum plane.

- The orientation of the holes will be relative to datum axis A.

86

In Panel B the datum precedence is B primary and A secondary R.F.S.

- A moveable gage is required and no datum shift is permissible on datum feature A.

- Datum feature B will have 3-point contact with its datum plane.

- The orientation of the holes will be relative to datum plane B.

In Panel C, the datum precedence is, B primary and A secondary MMC.

- A fixed gage is allowed when simulating datum A which allows datum shift.

- Datum feature B will have a three point contact with its datum plane.

- The orientation of the holes will be relative to datum plane B.

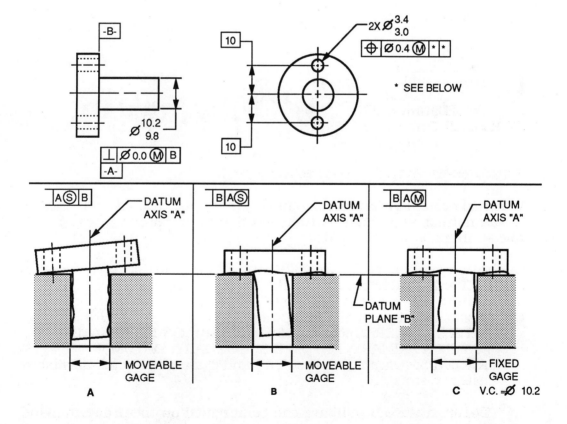

FIGURE 3-29 DATUM PRECEDENCE APPLICATION

SUMMARY

A summarization of datum concepts is shown in Figure 3-30.

CONCEPT	CAN BE APPLIED TO		IS APPLICABLE WHEN DATUM IS REFERENCED AT	
	FEATURE	FEATURE OF SIZE	RFS	MMC
DATUM SHIFT	NO	YES	NO	YES
DATUM TARGETS	YES	YES	YES	YES
Ⓜ MODIFIER	NO	YES	NO	YES
Ⓢ MODIFIER	YES*	YES	YES	NO

* IS AUTOMATIC - NEVER SHOWN

FIGURE 3-30 DATUM CONCEPTS SUMMARY

VOCABULARY WORDS

Datum
Datum Feature
Datum Plane
Primary Datum
Secondary Datum
Tertiary Datum
Datum Reference Frame
Datum Precedence
3-2-1 Rule
Datum Target
Datum Axis
Virtual Condition Datums
Datum Shift
Co-Datums

THOUGHT QUESTIONS

1.   Discuss the effects of adding this note to a drawing "UNLESS OTHERWISE SPECIFIED, ALL DIMENSIONS ARE RELATED TO DATUM A PRIMARY, DATUM B SECONDARY, DATUM C TERTIARY" (Datum features A, B, and C are all planar features).

2.   Do all parts require datums to be specified on the drawing? Why?

PROBLEMS AND QUESTIONS

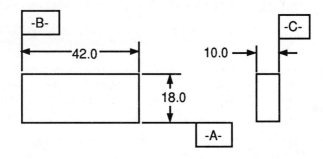

QUESTIONS 1-7 REFER TO THE FIGURE ABOVE

1. List the primary, secondary and tertiary datums _____ ,

    _____ , _____

2. Does datum plane A exist on the above part? _____

3. What is the angular relationship between datums A & B? _____

    _____

4. What is the angular relationship between datum features A & B?

    _____

5. If datum A is primary and datum B is secondary, how many points
    of contact are required for each? _____

    _____

6. Should the 10.0 dimension be checked from the datum reference
    frame? Why? _____

    _____

7. Is the 18.0 dimension a feature-of-size or is it related to datum
    plane A? Why? _____

    _____

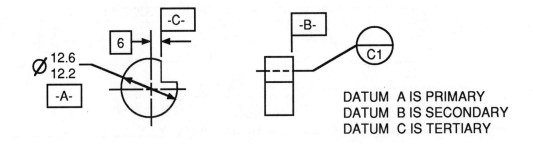

DATUM A IS PRIMARY
DATUM B IS SECONDARY
DATUM C IS TERTIARY

QUESTIONS 8-11 REFER TO THE FIGURE ABOVE

8. What is the minimum number of points of contact required to establish datum plane B? _____

9. Datum plane C is established by a _____ (line/point) datum target.

10. Are there enough dimensions shown to establish the datum target? _____

11. Does the 3-2-1 rule, for number of points of contact in the datum reference frame, apply for this part?_____

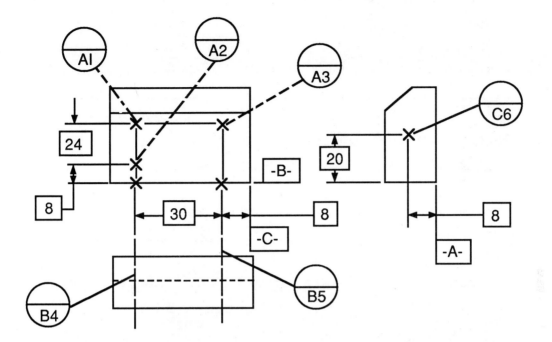

QUESTIONS 12 -18 REFER TO THE FIGURE ABOVE

12. _____ dimensions are used to locate the datum targets.

13. Datum targets exist on the _____ and are shown on the _____ .

14. Datum targets A1, A2, & A3 are _____ targets (points, lines, areas).

15. Datum targets B4 & B5 are _____ targets (points, lines, areas).

16. Is there enough information shown to establish a datum reference frame? _____

17. Which datum is primary? _____

18. Draw a cartoon gage for establishing the datum reference frame.

19. Can datum targets be used to establish a datum axis on a feature of size? _____

20. Can a basic dimension have a tolerance? Explain. _____

_____

21. Do basic dimensions exist on the part? Explain. _____

_____

22. What action should be taken when a part rocks on the primary datum plane? _____

23. Can datum shift occur with a part with planar datum features?

_____

24. Can datum shift occur when the datum feature is at MMC? _____

_____

25. Can a fixed gage be used to establish a datum axis referenced at MMC? _____

26. When a feature-of-size is referenced at MMC, the size of the gage is the _____ of the datum feature.

ACROSS

2. 2ND DATUM REFERENCED
6. SCARCE
9. NOT TRUE
10. SCHEMES
11. SCARCE
12. DATUM _____
14. ASSESS
19. RICKSHAW
20. KIND OF PAPER
23. ADDITIONAL TOLERANCE
24. 3RD DATUM REFERENCED
25. COOK IN A SKILLET
26. REFERENCE PLANE
27. PERUSE
29. RECEPTACLE
32. CONTACTS THE DATUM PLANE
34. MEASURE FOR BRICKS
35. SENIOR MEMBER
37. DATUM TARGET _____
42. TWELVE O'CLOCK
44. ARMY BUNK
45. "WHEN I WAS A _____ ..."
46. STAIR PART
47. TYPE OF DATUM TARGET
50. HIVE INSECT
51. DATUM SELECTION BASIS
54. INFORMATION
55. DATUM _____ FRAME

DOWN

1. EXPLOSION
2. OWN
3. DATUMS OF EQUAL IMPORTANCE
4. ENTRANCE
5. SHOOT THE BREEZE
6. OXIDATION
7. STRAY
8. ANNIVERSARY
13. DATUM PLANES ARE _____
15. EXIST
16. DATUM FEATURE EXISTS HERE
17. MIX
18. CAN BE USED TO ESTABLISH A DATUM
21. 1ST DATUM REFERENCED
22. BODY OF WATER
26. USED TO ESTABLISH A DATUM AXIS
28. PERFORMS IN A PLAY
30. NOT OCCUPIED
31. DIMENSIONS USED TO DEFINE DATUM TARGETS
33. DATUM MODIFIER
36. BUILDING PARCEL
38. FINIS
39. " ONE _____ THE MONEY ... "
40. PLOT
41. _____ DATUM
43. DATUM TARGET _____
44. WIRE
48. WHERE DATUMS EXIST
49. WATER AT LESS THAN 32 DEGREES
52. COOLING DEVISE
53. SHORT REST

CHAPTER **4**

ORIENTATION
CONTROLS

A trifle is often pregnant with high importance; the
prudent man neglects no circumstance.

Sophocles

INTRODUCTION

This chapter is about controlling the orientation (attitude) of part
features relative to each other. Every part feature must have some
orientation relative to the rest of the part. Sometimes a general note
like "Unless Otherwise Specified, All 90° Angles to be XX" will suffice.
In many cases, a direct orientation control, like parallelism, angularity,
or perpendicularity is required to meet the functional requirements of
the part.

GENERAL INFORMATION

When no orientation controls are specified on a drawing, the orientation (i.e. squareness, angularity, & parallelism) of part features is controlled by one of the following methods. Lines shown at right angles often have their tolerance controlled by an angular dimension with a tolerance, or a general note for angular tolerances on the drawing. Features which are shown parallel on a drawing are often controlled by the limits of the dimension locating the feature surfaces in conjunction with Rule #1. These are typical methods of dimensioning part features. Orientation controls become necessary when the type of controls mentioned above are inadequate or insufficiently accurate to satisfy part functional requirements.

Orientation Controls define the angularity, squareness, and parallelism of part features relative to one another. Orientation controls are sometimes referred to as attitude controls. There are three main orientation controls; perpendicularity, angularity, and parallelism. The symbols which designate these controls are shown in Figure 4-1. Orientation controls are considered "Related Feature Tolerances", which means that they must contain a datum reference in the feature control frame. The primary intent of these controls is to control the attitude of part features (and features-of-size) relative to other part features (or features-of-size). Other geometric controls using datum references also control attitude, but, as a secondary function.

PERPENDICULARITY	$\perp$
ANGULARITY	$\angle$
PARALLELISM	$/\!/$

FIGURE 4-1 ORIENTATION CONTROLS

ORIENTATION TOLERANCE ZONES

Orientation tolerance zones are total in value. This means that an axis, or centerplane, or all elements of a surface must fall within the tolerance zone specified by the orientation control. There are three types of orientation control tolerance zones. They are:

    Two parallel planes
    Two parallel lines
    Cylindrical

Note that when perpendicularity, angularity, and parallelism are applied to plane surfaces, the flatness of the surface is controlled to within the orientation tolerance zone. The tolerance zone in this application is two parallel planes within which all points of the considered surface must lie. Using logic, one can reason that if all points of the surface must fall within this zone to satisfy the orientation control, the flatness of the surface cannot be greater than this zone.

PERPENDICULARITY

Perpendicularity is the condition of a surface, or centerplane, or axis being exactly 90° to a datum. A *perpendicularity tolerance* is the amount which a surface, or axis, or centerplane is permitted to vary from being perpendicular to the datum.

Most perpendicularity applications fall into one of four types of general cases:

1. Perpendicularity applied to a surface or a planar feature-of-size.

2. Perpendicularity applied to a diameter (in one direction only).

3. Perpendicularity applied to the axis of a diameter.

4. Perpendicularity applied to a surface line element.

Figure 4-2 describes the four general types of perpendicularity applications.

CASE #	PERPENDICULARITY CONTROL	APPLIED TO A	SHAPE OF TOLERANCE ZONE	COMMMMENTS
1	⊥ 0.2 A	SURFACE	TWO PARALLEL PLANES	CONTROLS SURFACE ONLY
2	⊥ 0.2 A	HOLE	TWO PARALLEL PLANES	CONTROLS AXIS IN ONE VIEW ONLY
3	⊥ Ø 0.2 A	HOLE	CYLINDRICAL	CONTROLS AXIS IN ALL DIRECTIONS
4	⊥ 0.2 A	SURFACE*	TWO PARALLEL LINES	CONTROLS SURFACE LINE ELEMENTS

* WITH "EACH LINE ELEMENT" NOTE

FIGURE 4-2 PERPENDICULARITY APPLICATIONS

APPLICATIONS

Case 1 - Perpendicularity applied to a surface or a planar feature-of-size.

In this case, the perpendicularity control specifies a tolerance zone defined by two parallel planes perpendicular to a datum plane or axis within which the surface or median plane of the feature must lie. If the perpendicularity control is applied to a feature (planar surface), then it controls the attitude of the surface. See Figure 4-3. If the perpendicularity control applies to a feature-of-size, then it controls the attitude of the centerplane of the feature-of-size. See Figure 4-4.

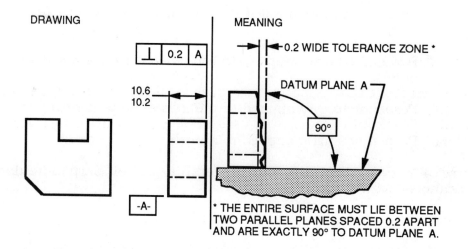

FIGURE 4-3 PERPENDICULARITY APPLIED TO A PLANAR SURFACE

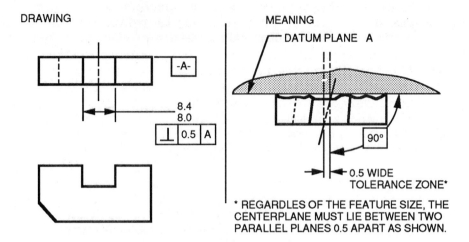

FIGURE 4-4 PERPENDICULARITY APPLIED TO A
SLOT-CENTERPLANE CONTROL

SUMMARY OF INFORMATION FROM FIGURES 4-3 & 4-4

- The perpendicularity control is applied to a feature in Figure 4-3
  and a feature-of-size in Figure 4-4. The location of the feature
  control symbol indicates whether it applies to a feature or a feature-
  of-size.

- Since the perpendicularity control applies to a feature in Figure 4-3,
  the following applies:

    - Rule #1 (applies to the form of the feature-of-size).
    - Rule #3 applies.
    - The flatness of the considered feature is limited by the
      perpendicularity tolerance zone.
    - The surface is controlled within two parallel planes.

- Since the perpendicularity control applies to a feature-of-size in
  Figure 4-4, the following applies:

    - Rule #1 is overridden and the form of the slot is controlled
      by a combination of the size tolerance and the
      perpendicularity tolerance.
    - Rule #3 applies.
    - The centerplane is controlled within two parallel planes.

- When inspecting these parts, two separate checks are required; one
  for size and one for the orientation control.

- Since the perpendicularity control in Figure 4-4 applies RFS, no
  bonus tolerance is permissible.

Case 2 - Perpendicularity applied to a diameter (in one direction only)

In this case, the perpendicularity control specifies a tolerance zone defined by two parallel planes perpendicular to a datum plane or axis within which the axis of the toleranced feature-of-size must lie.   See Figure 4-5.

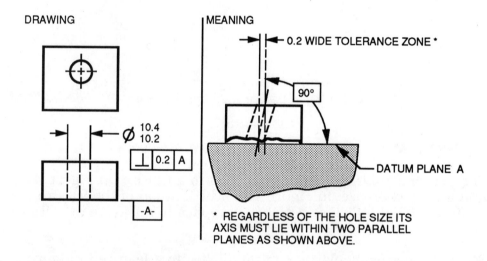

FIGURE 4-5 PERPENDICULARITY OF A FEATURE-OF-SIZE AXIS
IN ONE DIRECTION ONLY

SUMMARY OF INFORMATION FROM FIGURE 4-5

- The perpendicularity control is applied to a feature-of-size in Figure 4-5. The following applies.

    - The absence of a diameter symbol in the feature control frame indicates that the shape of the tolerance zone is two parallel planes.

    - Rule #3 applies.

    - When inspecting this part, two separate checks are required, one for the size of the hole, and one for the orientation of the hole.

100

Case 3 - Perpendicularity applied to the axis of a diameter.

In this case, the perpendicularity control specifies a cylindrical tolerance zone perpendicular to a datum plane or axis within which the axis of the considered feature must lie. (Note the diameter symbol must appear in the tolerance portion of the feature control frame.) See Figures 4-6 and 4-7.

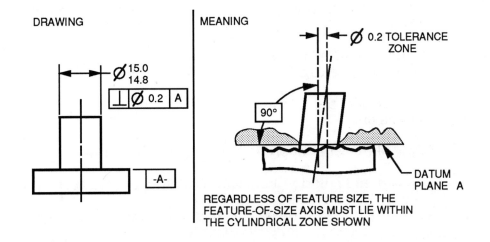

FIGURE 4-6 PERPENDICULARITY OF A FEATURE-OF-SIZE AXIS

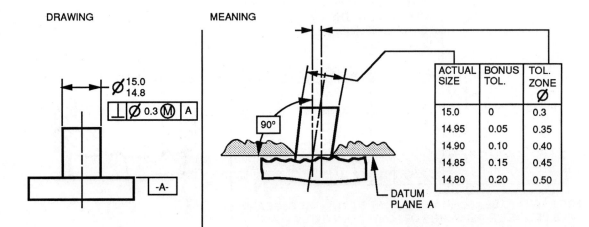

FIGURE 4-7 PERPENDICULARITY OF AN AXIS AT MMC

101

SUMMARY OF INFORMATION FROM FIGURES 4-6 & 4-7

- In Figure 4-6, the perpendicularity control applies regardless of feature size.

    - Rule #3 applies.
    - Bonus tolerance is not applicable.
    - The shape of the tolerance zone is cylindrical and controls the axis only.

- In Figure 4-7, the perpendicularity control applies at maximum material condition:

    - Bonus tolerance is applicable.
    - One fixed gage can be used to check perpendicularity of all parts.
    - The shape of the tolerance zone is cylindrical and controls the axis only.

- The following statements apply to both Figures 4-6 & 4-7:

    - The perpendicularity control is applied to a feature-of-size.
    - Rule #1 is overridden.
    - The straightness error of the axis of the shaft is limited to within the perpendicularity tolerance zone.
    - When inspecting, three separate checks are required; size, location, and orientation.

Figure 4-8 illustrates a fixed gage for checking perpendicularity at maximum material condition for the part shown in Figure 4-7.

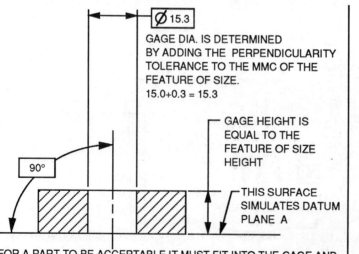
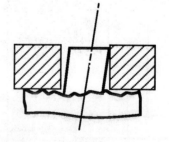

Ø 15.3

GAGE DIA. IS DETERMINED BY ADDING THE PERPENDICULARITY TOLERANCE TO THE MMC OF THE FEATURE OF SIZE.
15.0+0.3 = 15.3

GAGE HEIGHT IS EQUAL TO THE FEATURE OF SIZE HEIGHT

THIS SURFACE SIMULATES DATUM PLANE A

90°

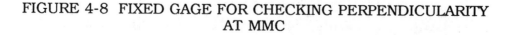
FOR A PART TO BE ACCEPTABLE IT MUST FIT INTO THE GAGE AND HAVE AT LEAST THREE POINTS OF CONTACT ON DATUM PLANE A

FIGURE 4-8 FIXED GAGE FOR CHECKING PERPENDICULARITY
AT MMC

When a design requires perfect orientation at maximum condition, it can be specified by using zero for the tolerance value in the feature control symbol. When it is necessary to limit the feature bonus effect to a maximum limit, an additional section can be added to the feature control symbol to indicate this requirement. Figure 4-9 shows an example of the above concept.

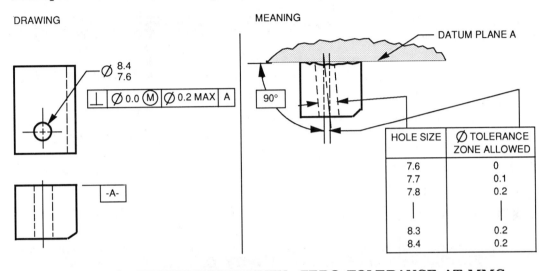

HOLE SIZE	Ø TOLERANCE ZONE ALLOWED
7.6	0
7.7	0.1
7.8	0.2
|	|
8.3	0.2
8.4	0.2

FIGURE 4-9   PERPENDICULARITY -ZERO TOLERANCE AT MMC
WITH A MAXIMUM SPECIFIED

Case 4 - Perpendicularity applied to a surface line element.

In this case, the perpendicularity control defines a tolerance zone of two parallel lines perpendicular to a datum plane or axis. Line element control is specified on a drawing by the addition of the phrase "each line element" or "each radial element". See Figures 4-10 & 4-11.

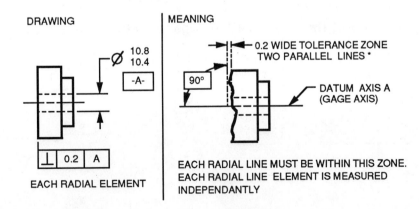

FIGURE 4-10   PERPENDICULARITY OF A LINE ELEMENT

103

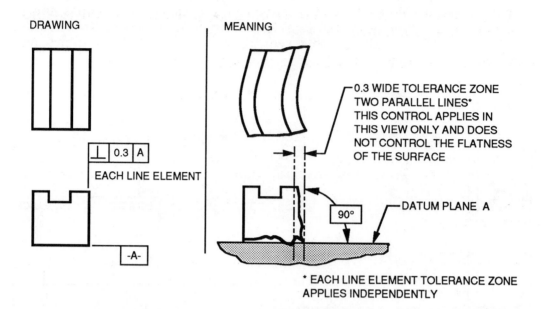

DRAWING

MEANING

0.3 WIDE TOLERANCE ZONE
TWO PARALLEL LINES*
THIS CONTROL APPLIES IN
THIS VIEW ONLY AND DOES
NOT CONTROL THE FLATNESS
OF THE SURFACE

⊥ | 0.3 | A

EACH LINE ELEMENT

90°

DATUM PLANE A

-A-

* EACH LINE ELEMENT TOLERANCE ZONE
APPLIES INDEPENDENTLY

FIGURE 4-11   PERPENDICULARITY OF A LINE ELEMENT

SUMMARY OF INFORMATION FROM FIGURES 4-10 & 4-11

The following statements apply to both figures:

- The perpendicularity control applies to the feature (line elements only).

- Since the control applies RFS, no bonus tolerance is permissible.

- Each line element check is independent.

- Rule #1 controls the form of the feature-of-size at MMC.

- When inspecting these parts, two separate checks are required; one for size and one for orientation.

LEGAL SPECIFICATION TEST

For a perpendicularity control to be a legal specification, it must satisfy the following conditions:

- A datum must be referenced in the feature control frame.

- If it is applied to a feature, no modifiers may be used for the tolerance value.

- The tolerance value specified must be a refinement of any other geometric tolerances that control the orientation of the toleranced feature (e.g. tolerance of position, total runout, profile of a surface).

A simple test for verifying if a perpendicularity control is legal is shown in Figure 4-12.

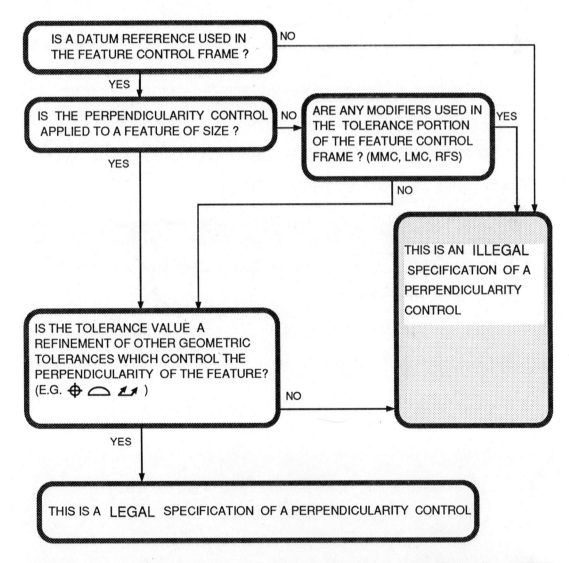

FIGURE 4-12 TEST FOR LEGAL SPECIFICATION - PERPENDICULARITY

ANGULARITY

Angularity is the condition of a surface, centerplane, or axis being exactly at a specified angle from a datum. An *angularity tolerance* is the amount which a surface, centerplane, or axis is permitted to vary from its specified exact angle. Angularity establishes a tolerance zone for a surface, centerplane, or axis which is a specified basic angle (other than 90°) from a datum plane or axis. An angularity tolerance zone is always two parallel planes. Most angularity tolerance applications fall into one of two types of cases.

APPLICATIONS

Case 1 - Angularity appled to a surface or a planar feature-of-size.

In this case, the angularity control specifies a tolerance zone defined by two parallel planes at the specified basic angle from a datum plane or axis within which the surface or centerplane of the considered feature must lie. See Figure 4-13.

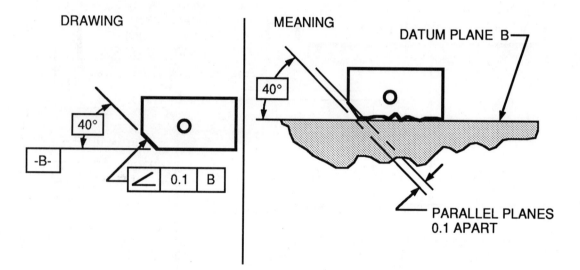

FIGURE 4-13 ANGULARITY APPLIED TO A SURFACE

Case 2 - Angularity applied to an axis.

In this case the angularity control specifies a tolerance zone defined by two parallel planes at the specified basic angle from a datum plane or axis within which the axis of the considered feature must lie. Note-this control limits the angularity of the axis in one plane. See Figure 4-14.

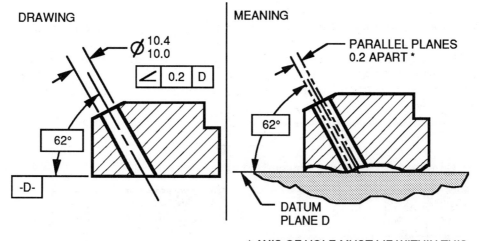

FIGURE 4-14 ANGULARITY APPLIED TO AN AXIS

SUMMARY OF INFORMATION FROM FIGURES 4-13 & 4-14

- In Figure 4-13, the following statements apply:

    - The angularity control is applied to a feature. The flatness error of the surface is limited to within the angularity tolerance zone.
    - A basic angle, from the datum reference plane (or axis) to the considered surface must be specified.
    - Rule #3 applies.

- In Figure 4-14, the following statements apply:

    - The angularity control is applied to a feature-of-size.
    - Angularity is controlled in one direction only.
    - Rule #3 applies.
    - A basic angle, from the datum reference plane (or axis) to the feature-of-size axis or enterplane must be specified.
    - Bonus tolerance is not applicable.

LEGAL SPECIFICATION TEST

For an angularity control to be a legal specification, it must satisfy the following conditions:

• A datum must be referenced in the feature control frame.

• If it is applied to a feature, no modifiers may be used for the tolerance value.

• A basic angle dimension must be specified from the toleranced feature to the datum plane.

• The tolerance value specified must be a refinement of any other geometric tolerances that control the angularity of the toleranced feature (e.g. tolerance of position, total runout, profile of a surface).

A simple test for verifying if an angularity control is legal is shown in Figure 4-15.

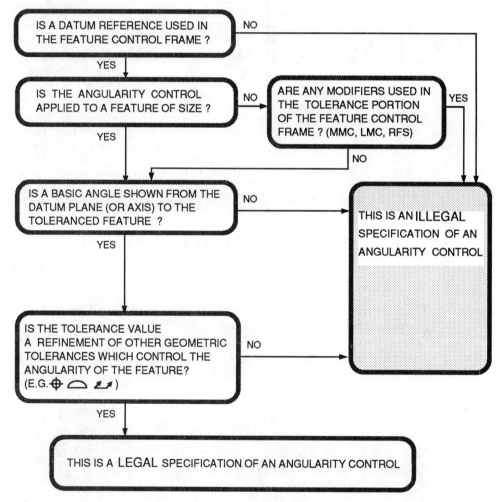

FIGURE 4-15 TEST FOR LEGAL SPECIFICATION - ANGULARITY

PARALLELISM

Parallelism is the condition where all points of a surface, centerplane, or axis, are an equal distance from a datum plane or datum axis. A *parallelism tolerance* is the amount which a surface, centerplane, or axis is permitted to vary from the parallel state. A parallelism control establishes a tolerance zone of two parallel planes or a cylinder within which all points of a controlled surface, centerplane or axis must lie. Most parallelism tolerance applications fall into one or two types of cases.

APPLICATIONS

Case 1 - Parallel planes as a tolerance zone.

In this case, the parallelism control specifies a tolerance zone defined by two planes parallel to a datum plane or datum axis. The distance between the planes is the tolerance value specified in the parallelism callout. All elements, line elements, or axes of the considered feature must lie within these planes. See Figures 4-16 & 4-17. In Figure 4-17, the parallelism control is applied to a hole, but the shape of the tolerance zone is two parallel planes. This is because there is no diameter symbol in the tolerance portion of the feature control frame.

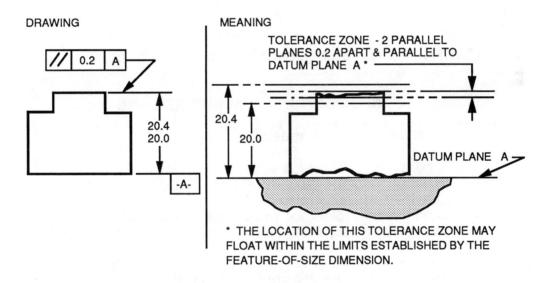

FIGURE 4-16 PARALLELISM APPLIED TO A SURFACE

109

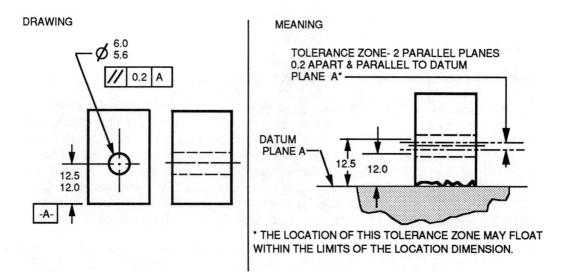

FIGURE 4-17 PARALLELISM APPLIED TO AN AXIS RFS

SUMMARY OF INFORMATION FROM FIGURES 4-16 & 4-17

- In Figure 4-16, the following statements apply:

    - The parallelism control is applied to a feature.
    - The flatness error of the surface is limited to within the parallelism tolerance zone.
    - The parallelism tolerance value is (and must be) less than the size tolerance. It is a refinement of the size dimension.
    - Rule #1 applies.
    - A datum reference is required.
    - Parallelism and size are checked separately.

- In Figure 4-17, the following statements apply:

    - The parallelism control is applied to a feature-of-size.
    - The parallelism tolerance is less than the location tolerance. It is a refinement of the feature location zone.
    - Rule #1 is overridden. The straightness of the axis is controlled to within the parallelism tolerance zone.
    - Rule #3 applies.

110

Case 2 - A cylinder as a tolerance zone.

In this case, a parallelism callout specifies a cylindrical tolerance zone is parallel to a datum axis within which the axis of the considered feature must lie. This cylindrical tolerance zone is designated by a diameter symbol in the tolerance portion of the feature control frame. The diameter of the cylindrical tolerance zone is equal to the tolerance value specified in the parallelism callout. See Figures 4-18 & 4-19.

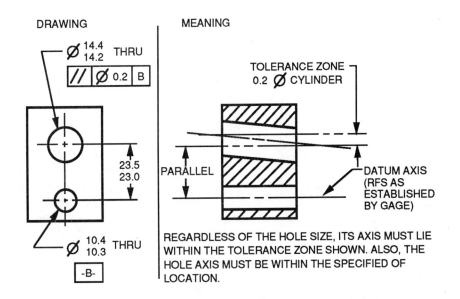

FIGURE 4-18 PARALLELISM OF A HOLE AXIS - RFS

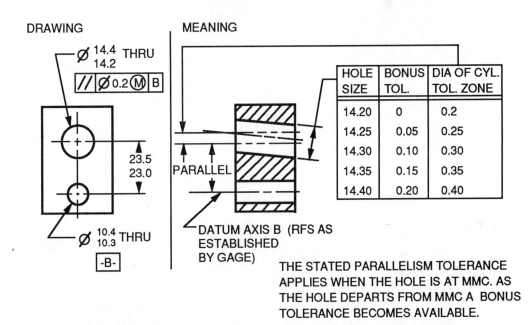

FIGURE 4-19 PARALLELISM OF A HOLE AXIS - MMC

111

SUMMARY OF INFORMATION FROM FIGURES 4-18 & 4-19

- In Figure 4-18, the bonus tolerance concept is not applicable.

- In Figure 4-19, the bonus tolerance concept is applicable.

- The following statements apply to both figures:

    - A datum reference is required.
    - The parallelism control is applied to the feature-of-size.
    - Rule #1 is overridden.
    - The straightness of the axis of the feature-of-size is limited to within the parallelism tolerance zone.
    - The parallelism tolerance value is less than the location tolerance. It is a refinement of the feature location zone.
    - When inspecting, three separate checks are required; size, location, and parallelism.

LEGAL SPECIFICATION TEST

For a parallelism control to be a legal specification, it must satisfy the following conditions:

- A datum must be referenced in the feature control frame.

- If it is applied to a feature, no modifiers may be used for the tolerance value.

- The tolerance value specified must be a refinement of any other geometric tolerances that control the parallelism of the toleranced feature (e.g. tolerance of position, profile of surface, total runout, size dimension).

A simple test for verifying if a parallelism control is legal is shown in Figure 4-20.

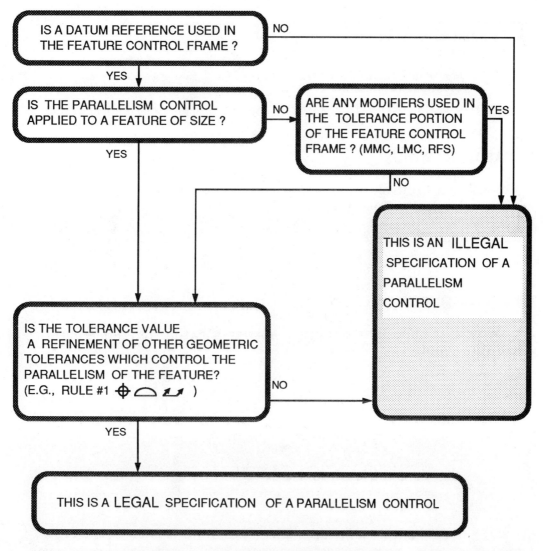

FIGURE 4-20 TEST FOR LEGAL SPECIFICATION - PARALLELISM

SUMMARY

A summarization of orientation control information is shown in Figure 4-21.

SYMBOL	DATUM REFERENCE REQUIRED OR PROPER	CAN BE APPLIED TO A		CAN AFFECT VIRTUAL CONDITION	CAN USE Ⓜ MODIFIER	CAN USE Ⓢ MODIFIER
		FEATURE	FEATURE OF SIZE			
⊥	YES	YES	YES	YES*	YES*	YES**
∠	YES	YES	YES	YES*	YES*	YES**
//	YES	YES	YES	YES*	YES*	YES**

* WHEN APPLIED TO A FEATURE OF SIZE
** AUTOMATIC PER RULE #3

FIGURE 4-21   SUMMARIZATION OF ORIENTATION CONTROLS

VOCABULARY WORDS

Orientation Control
Perpendicularity Tolerance
Angularity Tolerance
Parallelism Tolerance

THOUGHT QUESTION

1.   Why is it important to always specify a datum reference with an orientation control?

PROBLEMS AND QUESTIONS

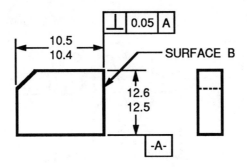

QUESTIONS 1-7 REFER TO THE FIGURE ABOVE

1. What is the shape and size of the tolerance zone for the orientation control? _____

2. Is the orientation control applied to a feature or a feature-of-size? _____

3. Is Rule #1 overridden for the 10.4-10.5 dimension? _____

4. What is the flatness of datum feature A limited to? _____

5. What is the flatness of surface B limited to? _____

6. If the 10.4-10.5 dimension was at MMC, what would the orientation of surface B be limited to? _____

7. Does the perpendicularity control apply at RFS or MMC? _____

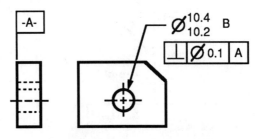

QUESTIONS 8-12   REFER TO THE FIGURE ABOVE

8. What is the virtual condition of hole B? _____

9. Describe the shape and size of the tolerance zone for hole B

_____

10. Does Rule #1 apply to the straightness of hole B?   Explain

_____

11. Would a 10.20 diameter pin always fit thru hole B?

_____

12. If the perpendicularity control was revised to ⟂|Ø 0.1 Ⓜ|A|
    what effect would this have on the inspection of the part?

_____

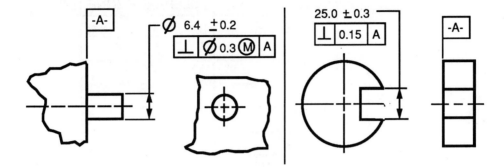

QUESTIONS 13-16 REFER TO THE FIGURE ABOVE

13. Fill in the chart (for each condition in the chart fill in the tolerance zone size).

ACTUAL FEATURE SIZE (DIA)	⊥ TOLERANCE ZONE DIAMETER
6.6 MMC	
6.5	
6.4	
6.3	
6.2 LMC	
6.1	

14. Draw and label the cartoon gage for checking the perpendicularity of this part.

15. Fill in the chart

ACTUAL FEATURE SIZE	WIDTH OF ⊥ TOLERANCE ZONE
24.7 MMC	
24.8	
24.9	
25.0	
25.1	
25.2	
25.3 LMC	

16. What is the shape and size of the tolerance zone for the perpendicularity callout? _____

_____

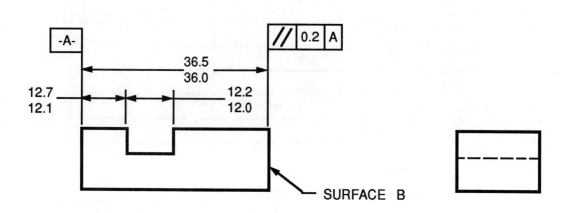

QUESTIONS 17-21   REFER TO THE FIGURE ABOVE

17. Describe the tolerance zone for the parallelism of surface "B."

_____

_____

18. What is the virtual condition of the width of the slot? _____

19. What is the virtual condition for the width of the block? _____

20. If the parallelism control on surface "B" was revised to $\boxed{\ /\!/\ \mid 0.1 \mid A\ }$ what would be the effect on the part? _____

_____

21. What controls the parallelism of the sides of the slot? _____

_____

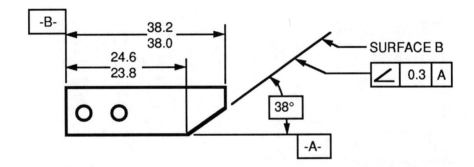

QUESTIONS 22-25 REFER TO THE FIGURE ABOVE

22. Describe the tolerance zone for the angularity callout ___38° ±0.15___

_____

23. The flatness of surface "B" is limited to _____

24. If the angularity callout was revised to $\boxed{\ \angle\ \mid 0.3 \mid A \mid B\ }$ what would be the effect on the tolerance zone? _____

_____

25. If the angularity callout was revised to $\boxed{\ \angle\ \mid 0.3 \mid B\ }$ what would be the effect on the tolerance zone? _____

_____

ACROSS

8. ANGULARITY, PERPENDICULARITY, PARALLELISM
11. AFTERNOON
13. TOLERANCE ZONE SHAPE
15. CHARGED ATOM
16. CAN BE CONTROLLED WITH PARALLELISM
20. RULE ABOUT RFS
21. CHILL
22. MAN'S NAME
23. DISENTANGLE
26. USED TO MODIFY A TOLERANCE ZONE
27. FARM ANIMAL
29. NUMBER OF ORIENTATION CONTROLS
30. MARKETING STRATEGY
32. CIRCULAR PART FEATURE
34. CONTROLS ANGLES
37. KING OF BEASTS
39. NATURE WALK
40. MORNING
41. LETTERS
42. BLACK
44. POSSIBLE WITH MMC MODIFIER
45. CAB
46. AUTO EFFICIENCY FIGURE

DOWN

1. MMC _____
2. ROMAN SEVEN
3. _____ GAGE
4. INSECT
5. "ONE _____ THE MONEY..."
6. BODY
7. NEGATIVE
9. CONTROLS IMPLIED 90 DEGREE ANGLES
10. HEALTH ESTABLISHMENT
11. TOLERANCE ZONE PART
12. USED WITH ANGULARITY
14. _____ 90 DEGREE ANGLE
17. A CRITICAL POINT
18. SOURCE OF ENERGY
19. BEWITCH
22. LARGEST CONTINENT
24. MATURE
25. TWO _____ PLANES
28. FALLS BEHIND
31. ORIENTATION CONTROLS ALWAYS USE ONE
33. TALK
35. SEIZE
36. BLOW UP
38. NEGATIVE WORD
40. HELP
43. TALENTED

120

CHAPTER **5**

LOCATION CONTROLS

A man should never be ashamed to own that he has been in the wrong, which is but saying in other words, that he is wiser today than yesterday.

Jonathan Swift

INTRODUCTION

This chapter explores the basics of location controls. Location controls are a very powerful engineering tool for defining, producing, and inspecting parts economically.

Much confusion exists today about location controls. This comes from a weak foundation in basic geometric dimensioning concepts like Rule #1, MMC concept, bonus, shift, etc. This chapter uses and discusses the basic principles necessary to understand location tolerances.

GENERAL INFORMATION

Location tolerances consist of; tolerance of position and concentricity. The symbols for these controls are shown in Figure 5-1. Location tolerances deal with features-of-size only. Therefore, it must be specified if they are to apply at MMC, RFS, or LMC.

Location tolerances are used to control three types of relationships. They are:

1. Center distance between features-of-size.

2. Location of a feature-of-size, or a group of features-of-size, relative to a datum or datums.

3. Coaxiality or symmetry of features-of-size.

REMEMBER

Location controls should always be applied to features-of-size.

A datum reference is required whenever a location tolerance is applied. (There is one exception to the above statement. It involves a tolerance of postion application and is described later in this chapter.)

REMEMBER

Location controls always require a datum reference.

POSTIONAL TOLERANCE

CONCENTRICITY

FIGURE 5-1 LOCATION CONTROL SYMBOLS

TOLERANCE OF POSITION

Tolerance of position is the most widely accepted location control used on engineering drawings today. This is because of its ability to describe the requirements of interchangeable components. One of the primary applications of tolerance of position is related to bolt hole pattern locations, because no other method so accurately describes the functional requirements of mating hole patterns. The comprehensive nature of tolerance of position will be seen in this chapter as its fundamental concepts are studied in greater detail.

ADVANTAGES OF TOLERANCE OF POSITION

A number of advantages can be cited from the use of tolerance of position. Although these advantages become more obvious as various aspects of tolerance of position are discussed in detail, some important benefits are listed to highlight the broad scope of this type of control.

- Round tolerance zones - 57% larger.

- Permits additional tolerances
    - Bonus
    - Shift

- Permits use of fixed gages

- Overcomes tolerance accumulation

- Protects part function

- Lowers production costs.

FUNDAMENTALS OF TOLERANCE OF POSITION

A tolerance of position is specified in a feature control frame by: the position symbol, a tolerance value, modifiers, and appropriate datum references. See Figure 5-2.

Two definitions regarding tolerance of position are:

> *True position* - The exact (perfect) location of a point, line, or plane (normally the center) of a feature-of-size in relationship with a datum reference frame and/or other features-of-size. Basic dimensions are used to establish the true position of features-of-size on drawings.

> *Tolerance of position* - The total permissible variation in the location of a feature-of-size about its true position.

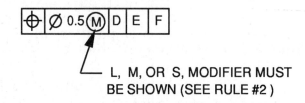

L, M, OR S, MODIFIER MUST
BE SHOWN (SEE RULE #2)

FIGURE 5-2 POSITIONAL TOLERANCE FEATURE CONTROL FRAME

REQUIREMENTS

Much of the confusion about tolerance of position comes from the numerous bad examples that exist on drawings. In this section, the BASIC REQUIREMENTS of a tolerance of position application are described. Drawing users may use this information to FIRST determine if a tolerance of position symbol is properly specified before attempting to interpret the application. If the tolerancing application does not fullfill these basic requirements, then the drawing must be challenged and corrected before proceeding.

There are four basic requirements in dimensioning systems using tolerance of position:

1. The tolerance of position must be applied to a feature-of-size.

2. Datum references are required.* Also, the datums must ensure that repeatable measurements of the considered feature can be made.

3. Basic dimensions are used to establish the true location of the features-of-size from the specified datums and between interrelated features-of-size.

4. LMC, MMC, or RFS modifiers must be specified in the feature control symbol as prescribed by Rule #2.

If any of these four conditions are not fullfilled, the tolerance of position specification is uninterpretable. See Figure 5-3 for examples of the above conditions.

* ANSI Y14.5M-1982 does include a special application of tolerance of position without a datum reference. This application is explained later in this chapter.

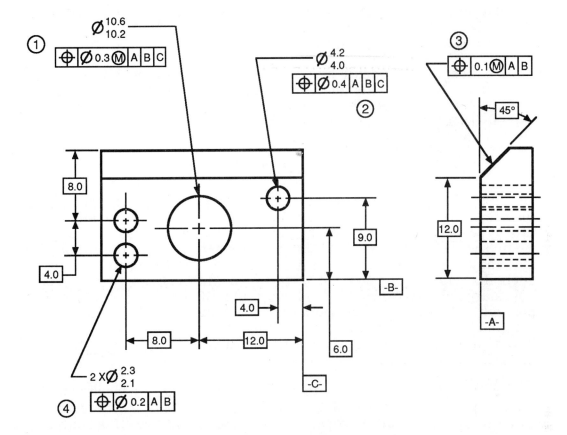

	POSITIONAL TOLERANCE	BASIC REQUIREMENT				COMMENTS
		1	2	3	4	
①	⊕ Ø 0.3 Ⓜ A B C	YES	YES	YES	YES	GOOD APPLICATION
②	⊕ Ø 0.4 A B C	YES	YES	YES	NO	DOES NOT SPECIFY IF TOL. APPLIES AT MMC OR RFS
③	⊕ 0.1 Ⓜ A B	NO	YES	YES	YES	POSITIONAL TOL. MUST BE APPLIED TO A FEATURE OF SIZE
④	⊕ 0.2 Ⓜ A B	YES	NO	NO	NO	NEEDS BASIC DIM FROM DATUM B NEEDS A TERTIARY DATUM TOL. AT MMC OR RFS?

FIGURE 5-3 FOUR REQUIREMENTS OF POSITIONAL
TOLERANCING DIMENSIONING

THEORY

Tolerance of position can be viewed in two ways:

1. As a boundary limiting the movement of the surface of a feature.

2. A tolerance zone limiting the movement of the axis of a feature.

Both concepts are useful and in most cases can be shown to be equivalent. However, the boundary concept is used extensively in this chapter because it represents functional requirements of mating parts and is the more flexible of the two systems.

THE BOUNDARY CONCEPT

To illustrate the boundary concept, let's examine the conditions resulting from a tolerance of position applied to a hole at MMC. In this type of application, the specified tolerance of position applies when the hole is at MMC (minimum diameter). The hole must be maintained within it's specified limits of size, and its location must be such that no surface element of the hole will be inside a theoretical boundary (also referred to as a gage pin diameter) located at true position. Also, it can be seen that the boundary size (9.8 dia.) is independent of the feature size. This provides an important insight into the relationship between size and tolerance of position. See Figure 5-4.

Since the boundary deals with the feature-of-size surface(s), it will always be three dimensional in nature. The height of the boundary is equal to the height of the considered feature-of-size. See Figure 5-5.

The diameter of a theoretical boundary for a tolerance of position (applied at MMC) of an *INTERNAL* feature-of-size is the MMC of the feature-of-size minus the tolerance of position value. See Figure 5-4. For an *EXTERNAL* feature-of-size, the boundary diameter is equal to the MMC of the feature-of-size plus the tolerance of position value. See Figure 5-6.

A tolerance of position is also an indirect orientation control. The attitude of the theoretical boundary produced from a tolerance of position is either perpendicular or parallel to the primary datum in the feature control frame. Since the surface of the feature-of-size is limited by this boundary, its attitude will be controlled by this boundary as well. See Figure 5-5.

Whenever a feature-of-size is controlled by a tolerance of position, Rule #1 is overridden. The straightness of the feature-of-size is also controlled by the boundary established by the tolerance of position. See Figure 5-5.

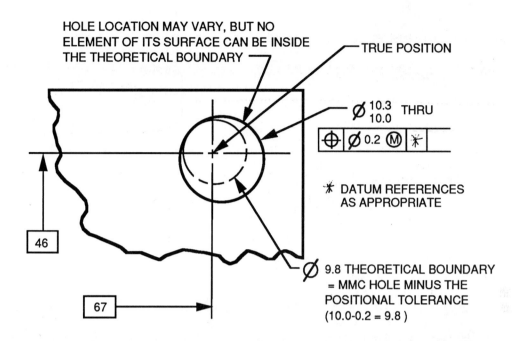

HOLE LOCATION MAY VARY, BUT NO
ELEMENT OF ITS SURFACE CAN BE INSIDE
THE THEORETICAL BOUNDARY

TRUE POSITION

∅ 10.3 / 10.0 THRU

⊕ ∅ 0.2 Ⓜ ✳

✳ DATUM REFERENCES
AS APPROPRIATE

46

67

∅ 9.8 THEORETICAL BOUNDARY
= MMC HOLE MINUS THE
POSITIONAL TOLERANCE
(10.0-0.2 = 9.8)

FIGURE 5-4 BOUNDARY CONCEPT FOR HOLE POSITION AT MMC

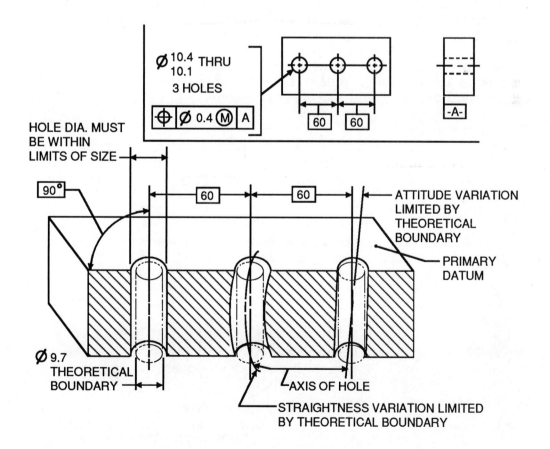

∅ 10.4 / 10.1 THRU

3 HOLES

⊕ ∅ 0.4 Ⓜ A

60   60

-A-

HOLE DIA. MUST
BE WITHIN
LIMITS OF SIZE

90°

60   60

ATTITUDE VARIATION
LIMITED BY
THEORETICAL
BOUNDARY

PRIMARY
DATUM

∅ 9.7
THEORETICAL
BOUNDARY

AXIS OF HOLE

STRAIGHTNESS VARIATION LIMITED
BY THEORETICAL BOUNDARY

FIGURE 5-5 POSITIONAL TOLERANCING BOUNDARY CONCEPTS

127

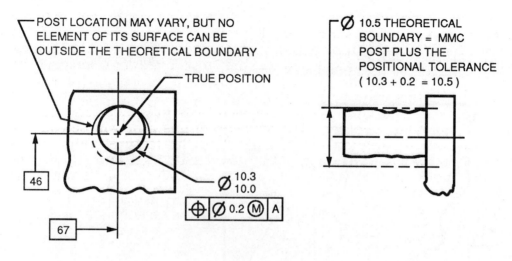

FIGURE 5-6
BOUNDARY CONCEPT FOR AN EXTERNAL FEATURE-OF-SIZE

THE AXIS CONCEPT

To illustrate a tolerance of position tolerance zone in terms of controlling the axis of a feature-of-size, let's examine the conditions resulting from a tolerance of position applied to a hole at MMC. In this type of application, when the hole is at MMC (its smallest diameter), its axis must fall within a cylindrical tolerance zone which is located at true position. See Figure 5-7. The diameter of this zone is equal to the tolerance of position value. This tolerance zone also defines the limits of the attitude of the axis of the hole at MMC in relation to the datum surface. The straightness of the hole at MMC is also limited by the cylindrical tolerance zone.

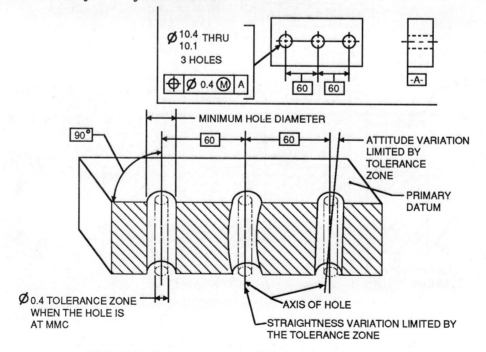

FIGURE 5-7 AXIS MOVEMENT CONCEPT

As stated earlier, the boundary and the axis concepts can in most cases be shown to be equivalent. Figure 5-8 shows how to convert a tolerance of position from a boundary concept to an axis tolerance zone. This can be accomplished by studying the effects of moving the hole until it contacts the boundary in various directions. This causes the center of the hole to generate a diametral tolerance zone about its true position. This zone is the equivalent axis tolerance zone derived from the boundary concept. The diameter of this zone will be equal to the tolerance of position value controlling the feature. Figure 5-9 shows how equivalent axis and boundary tolerance zones limit the location of a hole. Note; the total amount the hole can be from its true position is the same in both cases.

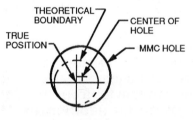

FIGURE 5-8 CONVERSION FROM A BOUNDARY
TO AN AXIS TOLERANCE ZONE

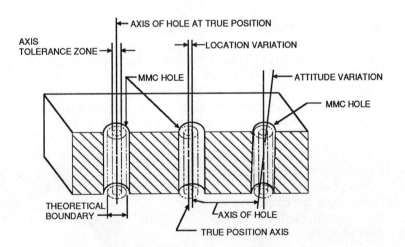

FIGURE 5-9 EQUIVALENT CONTROL OF HOLES
BY BOUNDARY AND AXIS ANALYSIS

Both tolerance zone concepts are useful. Because the axis concept is the basis of many inspection techniques, and is the basis for expressing tolerance of position requirements, it is an important concept in the understanding of tolerance of position theory. Since the boundary concept represents the true functional requirements of the mating part features, it is also an important concept in the understanding of tolerance of position theory.

APPLICATIONS (MMC)

A tolerance of position applied to a feature-of-size whose primary function is assembly, is usually applied at MMC. The MMC modifier is shown in the tolerance value portion of the feature control frame. When the feature-of-size is produced at MMC, and located at its extreme position, it is in its worst case (for assembly). (This is referred to as its virtual condition.) As the feature-of-size departs from MMC, an additional location tolerance, a Bonus tolerance, becomes available.

In other words, when a part is produced at its maximum material condition, it is the most demanding case for assembly (and no bonus is available). As the part feature-of-size departs from its MMC, its location can also vary by the same amount (in addition to its stated tolerance of position) and it will still assemble. This extra location tolerance, due to the considered feature departing from MMC, is called Bonus tolerance. See Figure 5-10.

REMEMBER

Whenever a tolerance of position is applied at MMC - a Bonus tolerance is available.

VERIFYING A POSITIONAL TOLERANCE WITH A FUNCTIONAL GAGE

A *functional gage* is a gage which verifies the functional requirements of part features. That is, if holes on a part are supposed to fit over studs of another part, the function of the holes would be to fit over the studs. To check the location of the holes, a functional gage which simulates the studs of the mating part, could be used. The following list highlights some of the benefits of functional gages:

- Economical to produce
- Gage can represents the worst case mating part
- Parts can be checked quickly
- No special skills required to "read" the gage, or interpret the acceptability of the results

When a functional gage has no moving parts, it is referred to as a fixed functional gage. The parts must fit with the gage to be acceptable.

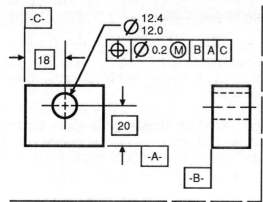

HOLE SIZE	LOCATION TOL ZONE	BONUS TOL	TOTAL LOCATION TOL ZONE DIA.
12.0	0.2	0	0.2 SEE A
12.1	0.2	0.1	0.3
12.2	0.2	0.2	0.4 SEE B
12.3	0.2	0.3	0.5
12.4	0.2	0.4	0.6 SEE C

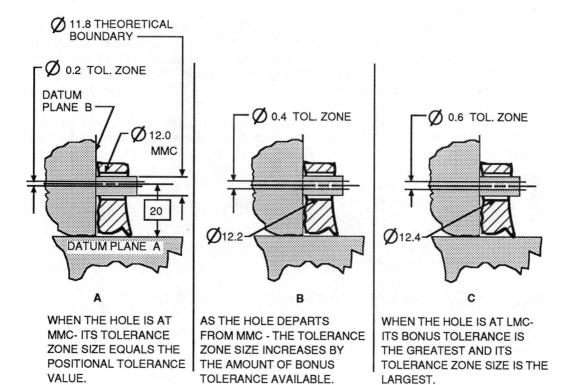

A

WHEN THE HOLE IS AT MMC- ITS TOLERANCE ZONE SIZE EQUALS THE POSITIONAL TOLERANCE VALUE.

B

AS THE HOLE DEPARTS FROM MMC - THE TOLERANCE ZONE SIZE INCREASES BY THE AMOUNT OF BONUS TOLERANCE AVAILABLE.

C

WHEN THE HOLE IS AT LMC- ITS BONUS TOLERANCE IS THE GREATEST AND ITS TOLERANCE ZONE SIZE IS THE LARGEST.

FIGURE 5-10 BONUS TOLERANCE CONCEPT
APPLIED TO POSITIONAL TOLERANCING

Functional gages are a common method for verifying tolerances of position. When verifying a tolerance of position, a functional gage establishes the datum planes (or axis) from the datum features, and, verifies that the considered part feature doesn't violate the theoretical boundary established by its tolerance of position.

Functional gages offer many advantages for checking parts dimensioned with tolerance of position, but, their use is not mandatory. Tolerance of position may also be verified with open inspection techniques.

Often it is desirable to analyze the extreme limits of a part in the design stage. This can be accomplished with a Cartoon gage. A *Cartoon gage* is a sketch of a functional gage. A designer uses a cartoon gage to verify the part extremes as an actual gage would.

The steps for determining the dimensions for a Cartoon gage are listed below:

- Determine the size of the gage feature. From the MMC of the toleranced feature, add or subtract (depending if it is an external or internal feature-of-size), the tolerance of position tolerance value to find the theoretical boundary which equals the size of the gage feature (hole, pin, slot, etc.).

- Establish datum surfaces (or axes) for datums referenced in the tolerance of position. Planar datum features are simulated by flat surfaces in the gage. If the datum features are features-of-size, and referenced at MMC, they will be simulated by a gage feature equal to the virtual size of the datum feature. (For a complete explanation of datum features-of-size refer to Chapter 3, page 77.)

- Locate gage features relative to their respective datums. The basic dimensions from the product drawing are used to locate the gage features relative to the datums.

This method is illustrated in Figures 5-11 & 5-12.

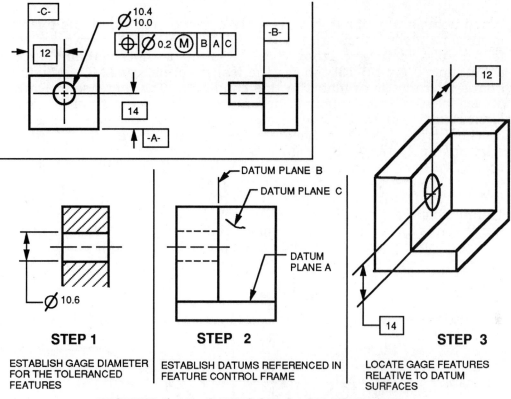

STEP 1

ESTABLISH GAGE DIAMETER FOR THE TOLERANCED FEATURES

STEP 2

ESTABLISH DATUMS REFERENCED IN FEATURE CONTROL FRAME

STEP 3

LOCATE GAGE FEATURES RELATIVE TO DATUM SURFACES

FIGURE 5-11   CARTOON GAGE EXAMPLE

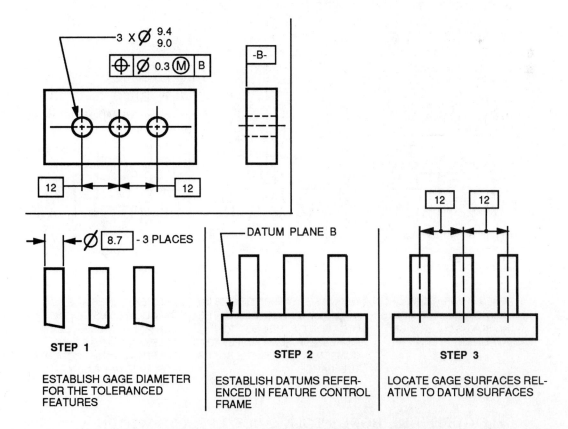

STEP 1

ESTABLISH GAGE DIAMETER FOR THE TOLERANCED FEATURES

STEP 2

ESTABLISH DATUMS REFERENCED IN FEATURE CONTROL FRAME

STEP 3

LOCATE GAGE SURFACES RELATIVE TO DATUM SURFACES

FIGURE 5-12   CARTOON GAGE EXAMPLE

133

When using a functional gage to analyze part extremes, bonus tolerance is automatically taken into account. Figure 5-13A shows an MMC feature-of-size in a cartoon gage. Bonus is not available in this condition. Figure 5-13B shows the feature-of-size at LMC. Maximum bonus tolerance is available in this condition. The increase in feature-of-size location, due to bonus, becomes a result of the smaller feature-of-size having more freedom to move within the fixed gage. The effect of the bonus tolerance becomes an automatic result of the part relationship to the gage. Figure 5-13C illustrates how part extremes can be calculated from gage and part dimensions. For example, let's say it is desired to find the maximum distance between the axis of the considered feature-of-size (the 10.0.-10.4 dia) and datum feature A. To find this maximum distance, the part is positioned in the gage with the features as far apart as possible. Gage and or part dimensions are added or subtracted to calculate the desired part condition.

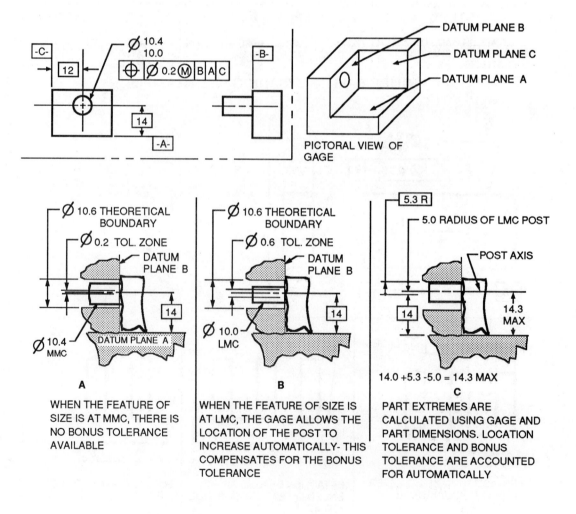

FIGURE 5-13 BONUS TOLERANCE EFFECTS WITHIN A
FUNCTIONAL GAGE

134

When a feature-of-size is used as a datum reference in a tolerance of position feature control frame, it must be specified if it applies at LMC, MMC, or RFS. (See Rule #2, Chapter 1.)

If a datum is referenced at MMC, the gage is a fixed size and a datum shift tolerance is possible. This is important to the part manufacturer because datum shift allows additional location tolerances. The amount of datum shift is different for each part produced. Datum shift, similar to bonus, is dependent upon the amount the feature-of-size departs from MMC.

REMEMBER

When an MMC modifier is used in the tolerance portion of a feature control frame - a bonus tolerance is available.

When an MMC modifier is used in the datum portion of a feature control frame - a shift tolerance is available.

If a functional gage is used to define the part extremes, bonus and datum shift are automatically accounted for in the part and gage analysis. See Figure 5-14.

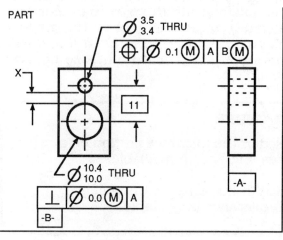

PART

$\emptyset\ \begin{smallmatrix} 3.5 \\ 3.4 \end{smallmatrix}$ THRU

| ⊕ | $\emptyset$ 0.1 Ⓜ | A | B Ⓜ |

X

11

$\emptyset\ \begin{smallmatrix} 10.4 \\ 10.0 \end{smallmatrix}$ THRU

| ⊥ | $\emptyset$ 0.0 Ⓜ | A |

-B-

-A-

GAGE

DATUM A

$\emptyset$ 3.3

11

✳

$\emptyset$ 10.0

✳ DATUM FEATURE SIMULATED AT VIRTUAL CONDITION

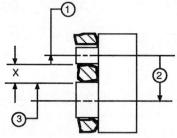

MAXMIUM X DISTANCE

① - 1.65 GAGE RADIUS
② +11.00 GAGE PIN LOCATION
③ - 5.0 GAGE RADIUS
_____
4.35

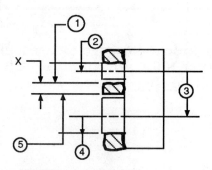

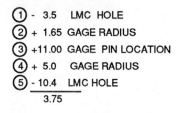

MINIMUM X DISTANCE

① - 3.5 LMC HOLE
② + 1.65 GAGE RADIUS
③ +11.00 GAGE PIN LOCATION
④ + 5.0 GAGE RADIUS
⑤ - 10.4 LMC HOLE
_____
3.75

FIGURE 5-14 CALCULATING A PART DISTANCE USING A
FUNCTIONAL GAGE

When a datum is referenced regardless of feature size in a feature control symbol, it is treated different from an MMC datum. An RFS datum refers to a centerplane or axis for a datum. The gage expands or contracts on the datum feature to establish this axis. There is no movement allowed of the datum feature in the gage. Therefore, no datum shift is possible. Because of this, a datum at RFS is a more stringent requirement than a datum at MMC. In a tolerance of position callout, an RFS datum should be used only where the design requirements cannot be fulfilled with an MMC datum. Figure 5-15 is an example of an RFS datum application.

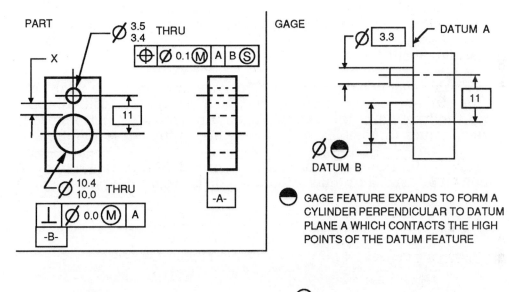

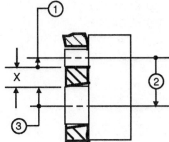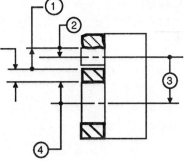

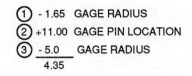

MAXMIUM X DISTANCE

① - 1.65 GAGE RADIUS
② +11.00 GAGE PIN LOCATION
③ - 5.0 GAGE RADIUS
  ‾‾‾‾‾‾
  4.35

MINIMUM X DISTANCE

① - 3.5 LMC HOLE
② + 1.65 GAGE RADIUS
③ +11.00 GAGE PIN LOCATION
④ - 5.2 GAGE RADIUS OF LMC HOLE
  ‾‾‾‾‾‾
  3.95

FIGURE 5-15 CALCULATING A PART DISTANCE
USING A FUNCTIONAL GAGE

137

APPLICATION (RFS)

A tolerance of position can be applied using the RFS modifier. This means that the tolerance of position value applies regardless of the feature size. This is a more stringent control than an MMC application. An RFS tolerance of position control should be used only where design requirements can not be met with an MMC modifier. There are three important concepts in using a tolerance of position with an RFS modifier.

a.   *No bonus tolerance is possible.* The tolerance of position value applies regardless of the feature size, the interrelationship between size and location no longer exists.

b.   *Axis tolerance zone control.* Tolerance of position applied RFS are easiest to visualize when thought of as an axis control. The diameter of the tolerance zone remains constant in size and is located about its true position.

c.   *Variable gaging must be used.* The gage must be movable to sense the size of the feature-of-size and the tolerance of position applies to the location of the axis of the feature-of-size. Open inspection techniques or complex variable gaging equipment is required.

In Figure 5-16, a tolerance of position applied RFS is illustrated. An example of a "max" and "min" calculated distance on the part is shown.

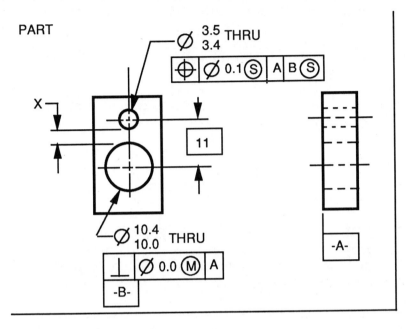

PART

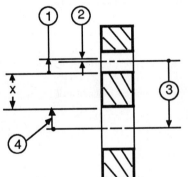

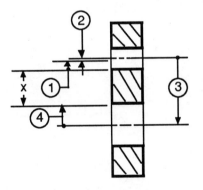

MAXMIUM DISTANCE X

① - 1.7   HOLE RADIUS

② + 0.05 TOL. ZONE RADIUS

③ + 11.00 BASIC LOC.

④ - <u>5.00</u>   HOLE RADIUS
     4.35

MINIMUM DISTANCE X

① - 1.75 HOLE RADIUS

② - 0.05 TOL. ZONE RADIUS

③ + 11.00 BASIC LOC.

④ - <u>5.2</u>   HOLE RADIUS
     4.00

FIGURE 5-16   CALCULATING A PART DISTANCE RFS APPLICATION

APPLICATION (LMC)

When the LMC modifier is applied to a tolerance of position, the specified tolerance value applies when the feature is at LMC. All of the principles for tolerance of position at MMC are reversed. The bonus is zero when the feature-of-size is at LMC and maximum when the feature-of-size is at MMC.

This modifier is used to control the minimum wall thickness of a part or to control minimum machine stock on a casting. See Figure 5-17.

Verification of a tolerance of position using the LMC modifier cannot be achieved with a fixed gage. Open inspection or variable gaging equipment is required.

DRAWING

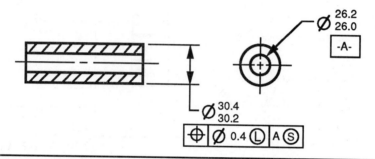

MEANING WITH PART AT LMC

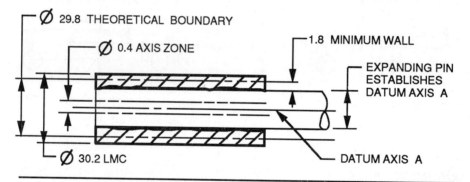

MEANING WITH PART AT MMC

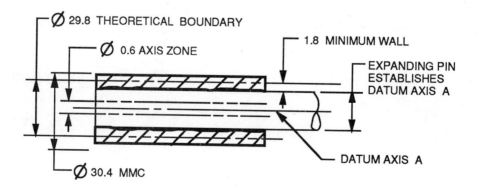

FIGURE 5-17 POSITIONAL TOLERANCING AT LMC

140

APPLICATION (PLANAR FEATURES-OF-SIZE)

The principles for tolerance of position of cylindrical features-of-size (such as holes and bosses) also apply to planar features-of-size (such as open end slots, tabs, etc.). In these types of applications, the tolerance of position value represents the distance between two parallel planes. The diameter symbol is omitted from the feature control frame, which indicates the tolerance zone is two parallel planes. See Figure 5-18.

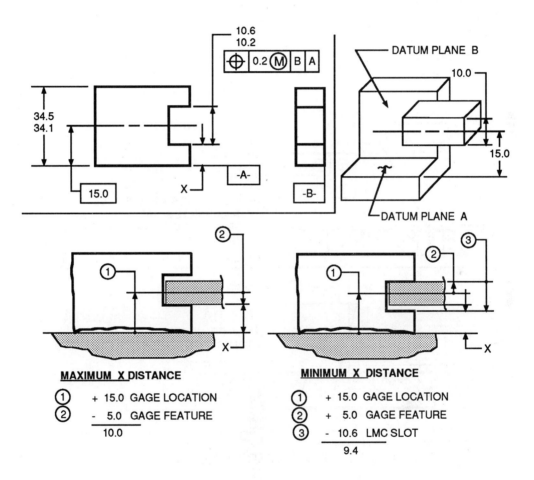

**FIGURE 5-18 POSITIONAL TOLERANCING OF A
PLANAR FEATURE OF SIZE**

APPLICATION (PROJECTED TOLERANCE ZONES)

When a projected tolerance zone modifier is used in conjunction with a tolerance of position, it means that the theoretical tolerance zone is projected above or below the part as indicated by the callout. The height of the projected zone is specified in the box below the feature control frame. Verification of the considered feature to its projected tolerance will predict whether the parts will assemble satisfactorily or not. See Figure 5-19. A projected tolerance zone is commonly used when the variation in perpendicularity of the threaded or press fit holes may cause the fasteners to interfere with mating parts. See Figure 5-20.

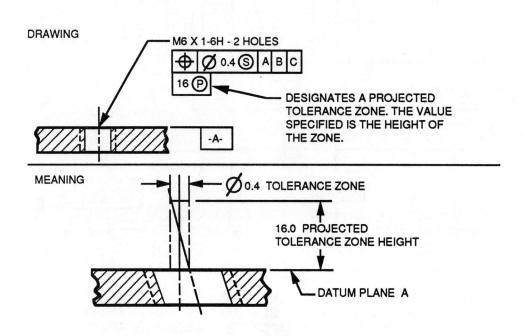

FIGURE 5-19 PROJECTED TOLERANCE ZONE

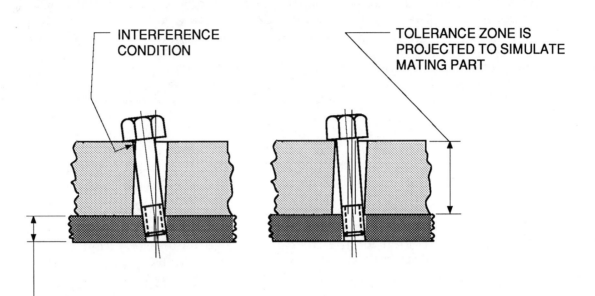

INTERFERENCE CONDITION

TOLERANCE ZONE IS PROJECTED TO SIMULATE MATING PART

TOLERANCE ZONE IS EQUAL TO LENGTH OF THREADED HOLE

PART WITHOUT PROJECTED TOLERANCE ZONE MODIFIER

PART WITH PROJECTED TOLERANCE ZONE (OF SAME TOLERANCE VALUE)

FIGURE 5-20 PROJECTED TOLERANCE ZONE

APPLICATIONS (COAXIAL FEATURES-OF-SIZE)

Coaxial features-of-size - Are two (or more) features-of-size, normally diameters, whose axes are coincident.

When it is desirable to allow the features-of-size to vary from a coaxial condition, the amount of variation may be specified by one of three geometric controls: concentricity, runout, or a tolerance of position. Generally, tolerance of position is the most liberal and least expensive coaxial control, which is why it is a common method for controlling the location of coaxial features-of-size. Runout is a tighter and more costly control. Its parameters will be discussed in Chapter 6. Concentricity is the most stringent, difficult to verify, and expensive control. Its parameters will be discussed later in this chapter. Functional design requirements determine the most appropriate geometric tolerance in an application involving coaxial features-of-size.

Whenever a tolerance of position is applied to coaxial features-of-size, a basic dimension of zero is implied between the feature axes. Also Rule #1 is overridden for the toleranced features-of-size. Figure 5-21 illustrates a part with a tolerance of position applied to coaxial features-of-size.

A tolerance of position can be used to control the location of coaxial holes. See Figure 5-22. In addition to controlling location, tolerance of position also controls the alignment of two or more coaxial holes. When a positional tolerance is used to control the alignment of coaxial features-of-size which are located with another tolerance of position (to a datum), no datum reference is required in the tolerance of position of the alignment callout. See Figure 5-22.

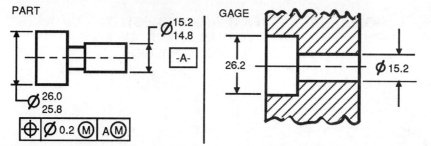

VARIOUS CONDITIONS OF PART IN GAGE

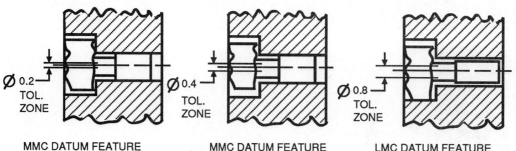

FIGURE 5-21 POSITIONAL TOLERANCE APPLIED TO
COAXIAL FEATURES-OF-SIZE

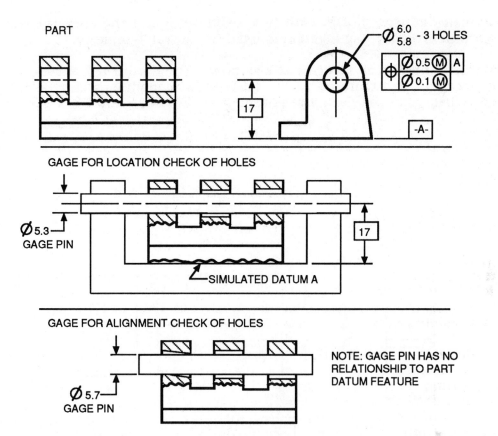

FIGURE 5-22   POSITIONAL TOLERANCING APPLIED TO
COAXIAL FEATURES-OF-SIZE

145

APPLICATION (SYMMETRY)

Symmetry is when a part feature is centered about the centerplane of a datum. Tolerance of position is used to control symmetry.

Symmetry applies to features-of-size only. The datum references must also contain a feature-of-size. MMC or RFS modifiers can be used depending upon the design requirements. See Figures 5-23 & 5-24

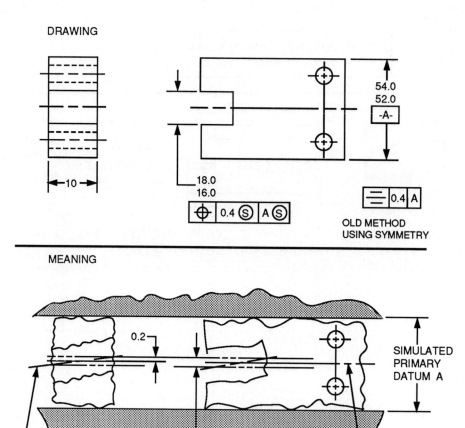

FIGURE 5-23   USING TOLERANCE OF POSITION TO CONTROL
SYMMETRY OF A SLOT (RFS)

146

DRAWING

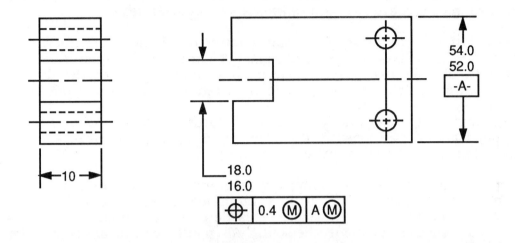

MEANING

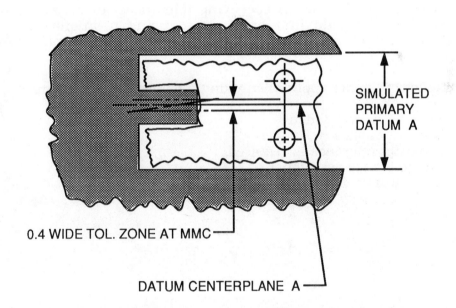

0.4 WIDE TOL. ZONE AT MMC

DATUM CENTERPLANE A

FIGURE 5-24   USING TOLERANCE OF POSITION TO CONTROL
SYMMETRY OF A SLOT (MMC)

CONCENTRICITY

A concentricity tolerance is indicated by the concentricity symbol, a tolerance value, and an appropriate datum reference placed in a feature control frame. See Figure 5-23.

Concentricity - The condition where the axis of a cylinder, cone, square, hex, etc. are common to the axis of a datum feature.

Concentricity Tolerance - The total amount of allowable variation of a feature-of-size axis to a datum axis. A concentricity tolerance is a cylindrical tolerance zone, whose axis is coincident with the datum axis, within which the axis of the considered feature-of-size must lie.

A concentricity tolerance zone and its datum reference can only be applied on an RFS basis. The size tolerance of a feature-of-size is independent of the concentricity tolerance.

The measurement of a concentricity tolerance requires the axis of the considered feature-of-size to be established by detailed analysis of circular elements of the surface. This determines a point of the axis for each circular element checked. All points of the axis must lie within the concentricity tolerance zone. Since irregularities in the form of the feature being inspected make it difficult to establish the axis of the feature, concentricity specifications should be avoided whenever possible. When specifying tolerances for coaxial features, consideration should first be given to using positional or runout tolerances. For an illustration of a concentricity application, see Figure 5-25.

The following items apply when using a concentricity callout:

- Rule #1 is overridden
- Rule #3 applies
- A datum reference is required
- The tolerance zone must be RFS
- The datum references must be RFS

PART

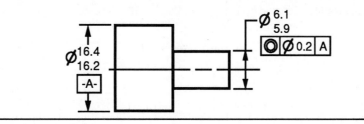

MEANING

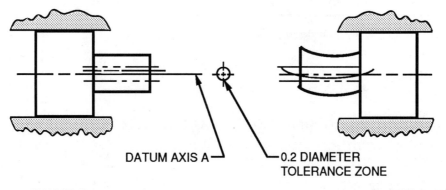

DATUM AXIS A —

— 0.2 DIAMETER
TOLERANCE ZONE

EXAMPLE #1   EXAMPLE #2

THE DERIVED AXIS OF THE CONSIDERED FEATURE MUST LIE WITHIN THE
CONCENTRICITY TOLERANCE ZONE. THIS AXIS IS ESTABLISHED BY ANALYSIS
OF THE SURFACE ELEMENTS OF THE CONSIDERED FEATURE.

FIGURE 5-25   CONCENTRICITY APPLICATION

SUMMARY

A summary of location control information is shown in Figure 5-26.

LOCATION CONTROL	DATUM REFERENCE REQUIRED	CAN BE APPLIED TO A		AFFECTS VIRTUAL CONDITION	CAN USE Ⓜ MODIFIER	CAN USE Ⓢ MODIFIER	OVERRIDE RULE #1
		FEATURE	FEATURE OF SIZE				
⊕	YES	NO	YES	YES	YES*	YES*	YES
◎	YES	NO	YES	YES	NO	YES**	YES

* Ⓜ OR Ⓢ MODIFIER MUST BE SHOWN PER RULE #2

** IS AUTOMATIC PER RULE #3

FIGURE 5-26 SUMMARIZATION OF LOCATION CONTROLS

VOCABULARY WORDS

True Position
Tolerance of Position
Functional Gage
Cartoon Gage
Coaxial Features-of-size
Concentricity

THOUGHT QUESTIONS

1. Discuss why a tolerance of position should not be used to control the location of a radius.

2. What are some of the advantages of using a zero tolerance at MMC to control the location of part features-of-size?

3. Should a tolerance of position be used to dimension the location of non-critical features-of-size?

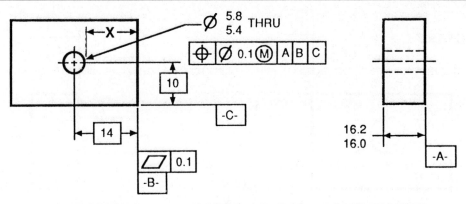

QUESTIONS 1-11 REFER TO THE ABOVE FIGURE

1. Which datum is primary? _____

2. What is the shape of the tolerance zone for the location of the hole? _____

3. What is the diameter of the gage pin for checking the location of the hole? _____

4. What is the attitude of the gage pin? _____

5. What is the location of the gage pin with respect to datums B & C? _____

6. What is the virtual condition of the hole? _____

7. How much bonus tolerance is associated with the tolerance of position callout? _____

8. What is the maximum distance "X" can be? _____

9. What is the minimum distance "X" can be? _____

10. What is the tolerance zone diameter for the axis of the hole when the hole is at LMC? _____

11. Draw a cartoon gage for checking the hole location.

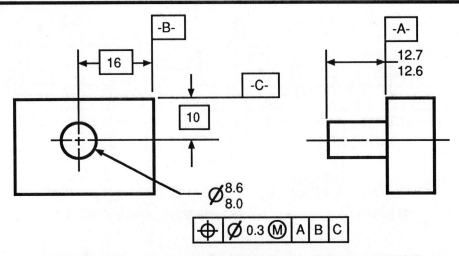

QUESTIONS 12-18 REFER TO THE ABOVE FIGURE

12. What is the MMC of the post? _____

13. What is the shape of the tolerance zone for the post location?

_____

14. How much bonus tolerance is associated with the tolerance of position callout? _____

15. What is the straightness of the post limited to? _____

16. What is the virtual condition of the post? _____

17. Fill in the chart below.

POST SIZE	POSITIONAL TOLERANCE	BONUS TOLERANCE	TOTAL LOCATIONAL TOLERANCE
8.6			
8.5			
8.4			
8.3			
8.2			
8.1			
8.0			

18. Draw a cartoon gage for checking the post location.

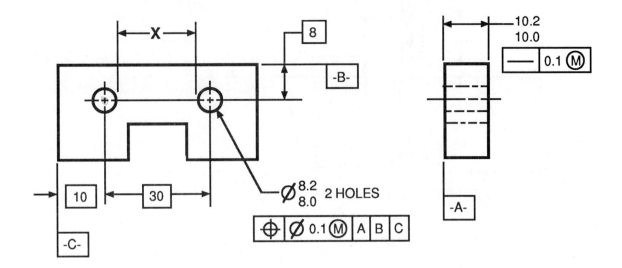

QUESTIONS 19-25 REFER TO THE ABOVE FIGURE

19. What is the MMC of the holes? _____

20. What is the diameter of the gage pin for checking the location of the holes? _____

21. What is the bonus tolerance associated with the tolerance of position callout? _____

22. What is the straightness of the holes limited to? _____

23. Draw a cartoon gage for checking the location of the holes on this part.

24. What is the maximum distance "X" can be on this part? _____

25. What is the minimum distance "X" can be on this part? _____

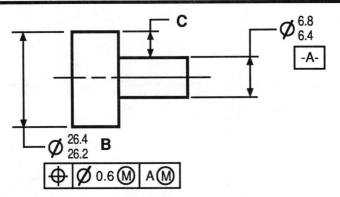

QUESTIONS 26-33 REFER TO THE ABOVE FIGURE

26. What is the size and shape of the tolerance zone for the location of diameter B? _____

    _____

27. How much bonus tolerance is associated with the tolerance of position callout? _____

28. What is the MMC of diameter B? _____

29. What is the MMC of datum feature A? _____

30. What is the virtual condition of diameter B? _____

31. What is the straightness of diameter B limited to?

    _____

32. Draw and dimension a cartoon gage for the tolerance of position callout.

33. If the straightness symbol [— | 0.3 Ⓜ] was applied to 6.8/6.4 dimension, what would be the maximum of distance C? _____ the minimum of distance C? _____

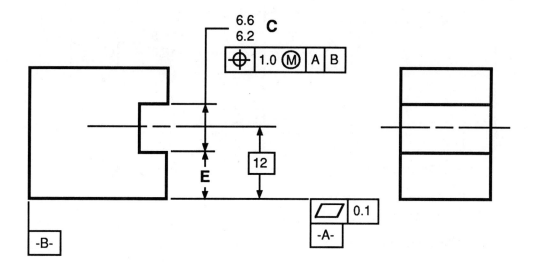

QUESTIONS 34-38 REFER TO THE ABOVE FIGURE

34. What is the shape and size of the tolerance zone for the location of slot C? _____

35. Draw a cartoon gage for checking the location of the slot.

36. What is the virtual condition of slot C? _____

37. What is the maximum distance E can be? _____

38. What is the minimum distance E can be? _____

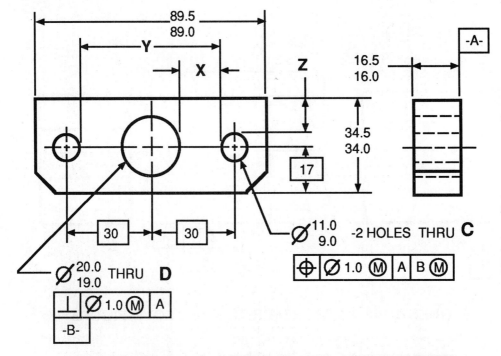

QUESTIONS 39-43 REFER TO THE ABOVE FIGURE

39. How much bonus tolerance is available for hole D? _____

40. How much shift tolerance is available when checking the location of holes C? _____

41. What is the virtual condition of hole D? _____

42. Draw a cartoon gage for checking the location of holes C.

43. Fill in the chart

DISTANCE	MAXIMUM	MINIMUM
X		
Y		
Z		

156

ACROSS

1. SEATING AREA IN RESTAURANT
4. LUNAR OR LEAP
7. MODIFIER FOR NO BONUS
8. A FUNCTIONAL TOLERANCE ZONE
11. BOMB SHELL THAT FAILS TO EXPLODE
12. L.A. PLAYERS
14. TOLERANCE ZONE SHAPE
16. A LOCATION CONTROL (3 WDS.)
19. SEIZE
25. REQUIRED FOR A LOCATION CONTROL
26. DIAMETER (3 WDS.)
27. INSIDE
29. TRIUMPH
32. PUNCTURE
33. HOG
35. _____ CONCEPT
36. EXTRA
37. PART MOVEMENT IN GAGE
39. AJAR
40. ANYTIME
43. ONE IN SPANISH
45. BANANA FEATURE
46. OUTLINE

DOWN

2. FORMER
3. RADIO BUFFS
5. OPPOSITE OF BEGINNING
6. WESTERN SHOW
7. RULE ABOUT TOLERANCE OF POSITION
9. _____ TOLERANCE
10. GUY
13. HEALTH ESTABLISHMENT
15. _____ CONTROL
17. AN AXIS TO AXIS CONTROL
18. TECHNICAL BOOK AUTHOR
20. VIEW
21. SAY FURTHER
22. MODIFIER TO PERMIT BONUS
23. GAGE WHICH MEASURES (2 WDS.)
24. GAGING WITH NO MOVING PARTS (2 WDS.)
26. _____ GAGE
28. DIMENSION WITH NO TOLERANCE
30. RASCAL
31. GAGE SKETCH
34. ADHESIVE
38. GREATER THAN 98.6 DEGREES
41. ASSESS
42. 2,000 lbs.
44. PREFIX INDICATING INVERSE SHAPE OR
    ATTACHMENT

CHAPTER 6

RUNOUT CONTROLS

You have little chance of producing a quality product from vague incomplete specifications.

INTRODUCTION

In this chapter runout controls are explored. Runout controls consist of Circular runout and Total runout.

GENERAL INFORMATION

This chapter discusses two types of runout controls, circular runout and total runout. The symbol for each is shown in Figure 6-1. There are examples and information unique to each control, but the following information applies to both types of runout controls. *Runout* is a composite control affecting both the form and location of a part feature relative to a datum axis. Whenever a runout control is specified, a datum reference is required.

REMEMBER

Runout is a composite control affecting both form and location relative to a datum axis.

Notes on older drawings which used terms like F.I.M. (Full Indicator Movement) T.I.R. (total Indicator Reading) and T.I.M. (Total Indicator Movement) were all similiar to runout tolerances.

CIRCULAR RUNOUT

TOTAL RUNOUT

FIGURE 6-1 RUNOUT CONTROLS

A runout control can be applied to any diametral shape that is around the datum axis, or, to a surface that is perpendicular to the datum axis. Examples include, a diameter around the datum axis, a conical surface around the datum axis, or a surface perpendicular to the datum axis. Figure 6-2 illustrates these conditions.

REMEMBER

Runout must always be applied to a part feature which is around the datum axis.

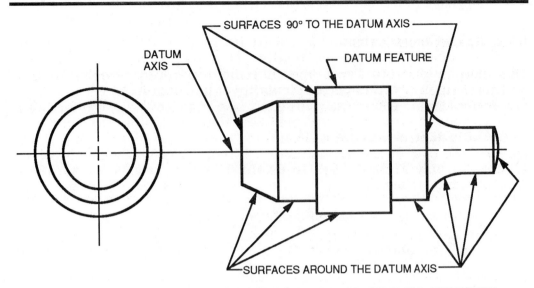

FIGURE 6-2 TYPES OF FEATURES APPLICABLE TO RUNOUT

A runout tolerance value specified in a feature control symbol indicates the maximum permissible indicator reading (or gage travel) of the considered feature when the part is rotated 360° about its datum axis. See Figure 6-3.

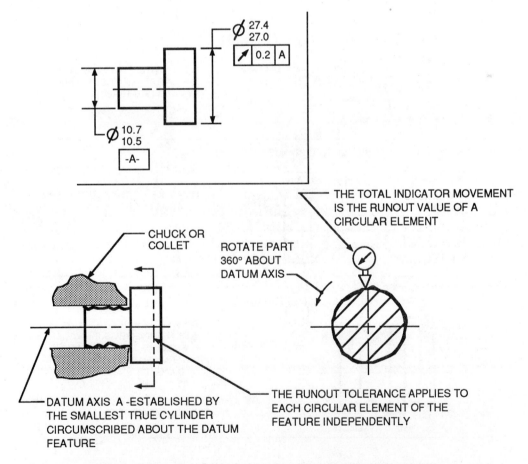

CHUCK OR COLLET

ROTATE PART 360° ABOUT DATUM AXIS

THE TOTAL INDICATOR MOVEMENT IS THE RUNOUT VALUE OF A CIRCULAR ELEMENT

DATUM AXIS A -ESTABLISHED BY THE SMALLEST TRUE CYLINDER CIRCUMSCRIBED ABOUT THE DATUM FEATURE

THE RUNOUT TOLERANCE APPLIES TO EACH CIRCULAR ELEMENT OF THE FEATURE INDEPENDENTLY

FIGURE 6-3 RUNOUT TOLERANCE EXAMPLE

One of the most common applications of runout controls are the location of coaxial features relative to a datum axis. Other less common uses or (indirect benefits) include controlling circularity, wobble, angularity, and perpendicularity of part features.

ESTABLISHING A DATUM AXIS

There are only THREE ways to establish a datum axis for a runout specification. They are:

- A single diameter of sufficient length.

- Two coaxial diameters with sufficient distance between to create a single datum axis.

- A plane and a diameter at right angles.

Figure 6-4 illustrates the above conditions. Functional design requirements and part shape are considerations for selecting one of the above methods of establishing a datum axis. Usually the features chosen for the datum axis are the same features which locate the part (radially) in its assembly.

A SINGLE DIAMETER OF SUFFICIENT LENGTH

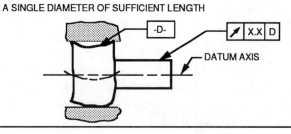

BOTH ATTITUDE AND POSITION OF THE DATUM AXIS ARE DETERMINED BY DATUM FEATURE D

TWO COAXIAL DIAMETERS A SUFFICIENT DISTANCE APART

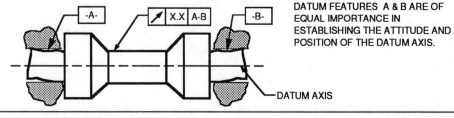

DATUM FEATURES A & B ARE OF EQUAL IMPORTANCE IN ESTABLISHING THE ATTITUDE AND POSITION OF THE DATUM AXIS.

A PLANE AND A DIAMETER

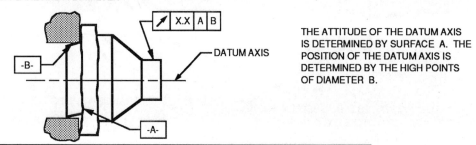

THE ATTITUDE OF THE DATUM AXIS IS DETERMINED BY SURFACE A. THE POSITION OF THE DATUM AXIS IS DETERMINED BY THE HIGH POINTS OF DIAMETER B.

FIGURE 6-4 THREE WAYS OF ESTABLISHING A DATUM AXIS

REMEMBER

There are only three ways to establish a datum axis for a runout control. They are:
- A single diameter of a sufficient length
- Two coaxial diameters with sufficient distance between
- A plane and a diameter at right angles

CIRCULAR RUNOUT

Circular Runout is a composite control affecting both the form and location of circular elements of a part feature. It is referred to as a composite control because it affects both form and location simultaneously.

Circular runout is frequently used to control the location of circular elements of a diameter. When applied to a diameter, it controls the circularity and coaxiality of the diameter to a datum axis. When applied to a surface 90° to the datum axis, it controls the attitude of the circular elements only. Circular runout applies to each circular element of a surface independent from one another.

The tolerance zone shape for circular runout is easily visualized. It can be thought of as two coaxial circles whose centers are located on the datum axis. The radial distance between these circles is equal to the runout tolerance value. See Figure 6-5A. The size of the larger circle is established by the radius of the surface element which is farthest from the datum axis. See Figure 6-5B. (Note the diameter must also meet its size requirements.)

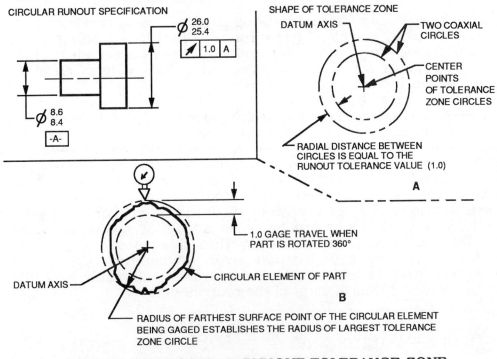

FIGURE 6-5 CIRCULAR RUNOUT TOLERANCE ZONE

163

> **REMEMBER**
>
> A circular runout control applies to each circular element independently.

When inspecting a diameter controlled with circular runout, a gage (dial indicator) is held perpendicular to the surface being checked. A sufficient number of circular elements should be checked to assure the entire diameter is within the specified tolerance. The number of elements checked is usually left to the judgement of the inspector. If necessary, the number and location of the gage readings can be specified on the drawing.

> **REMEMBER**
>
> When measuring runout, the gage is always held perpendicular to the surface being verified.

Circular runout is a composite control. It controls both the form and location of a feature simultaneously. Therefore, when verifying a runout specification, a gage contacts the actual part surface. Surface irregularities and roundness (form) automatically become part of the runout control gage reading. Figure 6-6 illustrates how form (circularity) affects runout.

In case 1, a circular element which is perfectly round and is coaxial with the datum axis is being gaged. As the part is rotated 360° about the datum axis, the gage reading (runout value) will be zero.

In case 2, a circular element is being gaged which is coaxial with the datum axis, but is not perfectly round. The gage reading (runout tolerance value) will detect this roundness error as a runout tolerance. The gage cannot separate a location error from a form error in a runout check.

In case 3, a circular element which is perfectly round is being gaged. In this example, all of the runout error will be a result of axis offset (eccentricity). The maximum allowable offset between the toleranced diameter axis and the datum axis is one half the runout tolerance value. This is because the runout tolerance zone can be converted into an axis tolerance zone, which is a cylinder, equal in diameter to the runout tolerance value and centered about the datum axis. The axis of the toleranced diameter can be offset above, or below (or in any direction from), the datum axis within this zone. This is the maximum possible axis location error. Whenever there is any roundness error in the considered feature, the allowable axis offset will be reduced by the amount of the roundness error.

164

REMEMBER

When a diameter is controlled with runout, the maximum its axis can be offset from the datum axis is half of the runout tolerance value.

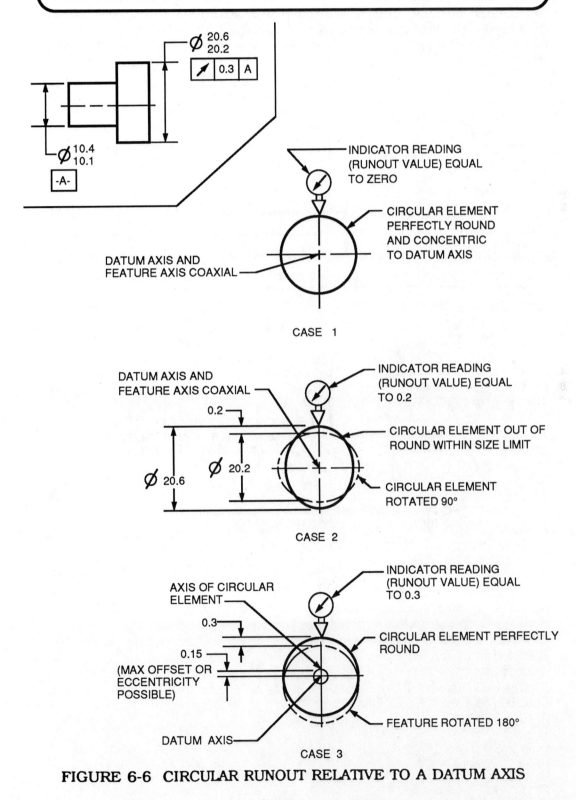

FIGURE 6-6 CIRCULAR RUNOUT RELATIVE TO A DATUM AXIS

When checking a surface (which is at a right angle to the datum axis) that is controlled with circular runout, the gage is held perpendicular to the surface being checked. The part (or gage) is rotated 360° and the variation of a circular element is the runout reading. See Figure 6-7.

Note, when circular runout is applied to a surface, as described above, it controls the line elements of the surface for attitude (wobble). In this type of application circular runout does not control the attitude of the surface.

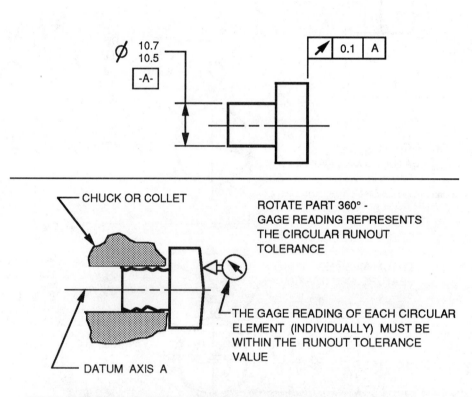

FIGURE 6-7 CIRCULAR RUNOUT APPLIED TO A SURFACE
PERPENDICULAR TO THE DATUM AXIS

LEGAL SPECIFICATION TEST

For the specification of a circular runout control to be legal, it must satisfy the following conditions:

- A datum must be referenced in the feature control frame.

- The datum reference must specify a proper datum axis.

- The control must be applied to a surface element which surrounds the datum axis.

- No modifiers may be used in the feature control frame.

A simple test for verifying if a circular runout control is properly specified is shown in Figure 6-8.

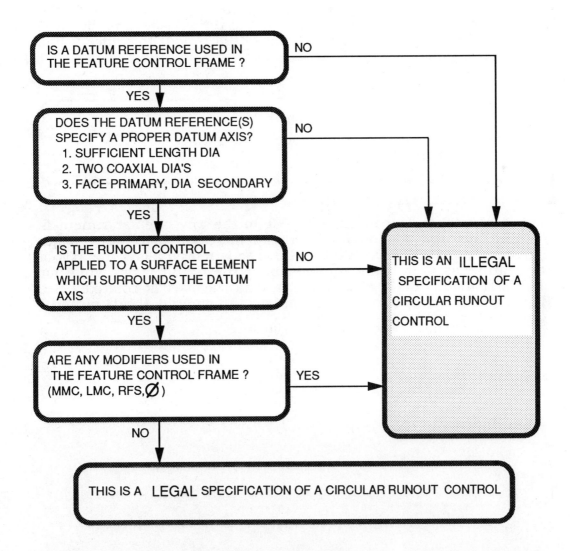

FIGURE 6-8   TEST FOR LEGAL SPECIFICATION - CIRCULAR RUNOUT

TOTAL RUNOUT

Total runout is a composite control affecting the form and location of all surface elements simultaneously.

When total runout is applied to a surface around a datum axis (such as a diameter or a cone), it controls cumulative variations of circularity, straightness, location, angularity, taper, and profile of the surface. When total runout is applied to a diameter, the tolerance zone is easily visualized. It consists of two coaxial cylinders whose centers are located on the datum axis. The radial distance between these cylinders is equal to the runout tolerance value. The size of the larger cylinder is established by the radius of the surface element which is farthest from the datum axis. See Figure 6-9. (Note the diameter must also meet its size requirements.)

When total runout is applied to a surface that is 90° to the datum axis, it controls variations of perpendicularity and flatness. In this case, the tolerance zone consists of two parallel planes (perpendicular to the datum axis) within which all the elements of the surface must lie. The distance between these planes is equal to the runout tolerance value. See Figure 6-10.

REMEMBER

A total runout specification applies to all surface elements simultaneously.

Gaging of total runout is quite similar to the way circular runout is gaged. They are both verified with an indicator gage. The major difference is that for total runout, the gage is moved along the axis of the feature as it is being rotated about the datum axis, while for circular runout, the gage is stationary as the part is rotated. See Figures 6-9 & 6-10 for examples of gaging total runout.

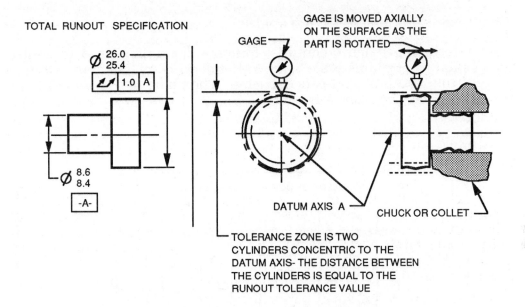

TOTAL RUNOUT SPECIFICATION

GAGE

GAGE IS MOVED AXIALLY ON THE SURFACE AS THE PART IS ROTATED

DATUM AXIS A

CHUCK OR COLLET

TOLERANCE ZONE IS TWO CYLINDERS CONCENTRIC TO THE DATUM AXIS- THE DISTANCE BETWEEN THE CYLINDERS IS EQUAL TO THE RUNOUT TOLERANCE VALUE

FIGURE 6-9 TOTAL RUNOUT TOLERANCE ZONE

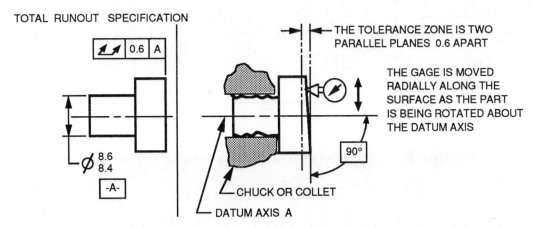

TOTAL RUNOUT SPECIFICATION

THE TOLERANCE ZONE IS TWO PARALLEL PLANES 0.6 APART

THE GAGE IS MOVED RADIALLY ALONG THE SURFACE AS THE PART IS BEING ROTATED ABOUT THE DATUM AXIS

90°

CHUCK OR COLLET

DATUM AXIS A

FIGURE 6-10 TOTAL RUNOUT APPLIED TO A SURFACE 90° TO THE DATUM AXIS

169

LEGAL SPECIFICATION TEST

This test is the same as the test for circular runout shown on page 167.

COMPARISON OF COAXIAL CONTROLS

Of the thirteen geometric controls, four are commonly used to control the coaxiality of diameters. They are: tolerance of position, circular runout, total runout, and concentricity. The chart in Figure 6-11 is a comparison of these controls.

COMPARISON OF COAXIAL CONTROLS					
CONTROL	BONUS	SHIFT	MANUFACTURING COSTS	GAGING COSTS	APPLICATION
⊕ ⌀ 0.0 Ⓜ A Ⓜ	YES	YES	LO ... HI	LO ... HI	ASSEMBLY
⊕ ⌀ 0.5 Ⓜ A Ⓜ	YES	YES			
⊕ ⌀ 0.5 Ⓜ A Ⓢ	YES	NO			
⊕ ⌀ 0.5 Ⓢ A Ⓢ	NO	NO			
↗ 0.5 A	NO	NO			ROTATING PARTS
⌰ 0.5 A	NO	NO			
◎ ⌀ 0.5 A	NO	NO			

FIGURE 6-11 COMPARISON OF COAXIAL CONTROLS

PART CALCULATIONS

The maximum (or minimum) distance between two external (or internal) diametral surfaces controlled with runout can be calculated quite easily. This applies to both circular and total runout. The following paragraphs describe how to make the calculations for external features. Once the process is understood for external features, the same technique can be applied when using internal features, or any combination thereof.

1. Identify start and end points

   • Draw an "X" on the two surfaces that are the beginning and end of the part distance being calculated. See Figure 6-12A.

2. Establish the sign convention

   • On the start point add a double ended arrow. Assign the direction towards the endpoint as "+". Assign the direction away from the endpoint as "-". See Figure 6-12A. Whenever a part distance is used in the "+" direction, it will be added to the calculation. Whenever a part distance is used in the "-" direction, it will be subtracted from the calculation.

3. Identiify a continuous path of known dimensional values.

   • From the start point to the end point, identify a path of part dimensions by drawing an arrow for each distance involved. See Figure 6-12B.

   • When using diameters it is the runout control that defines the distance between centerlines.

   • Using the sign convention show A "+" or "-" sign for each part distance.

4. Solve the maximum & minimum distance by adding the dimensions in the continuous path in step 3. See Figure 6-12C.

   • For maximum distance.

     - Use all maximum "+" values
     - Use all minimum "-" values
     - Add 1/2 of the runout tolerance values.

   • For the minimum part distance.

     - Use all minimum "+" values
     - Use all maximum "-" values
     - Subtract 1/2 of the runout tolerance values.

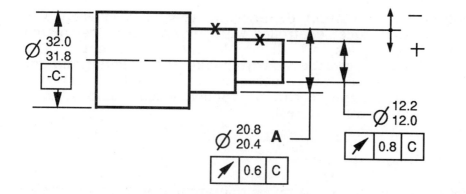

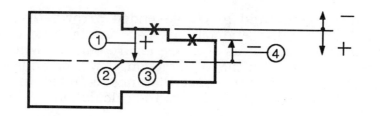

A

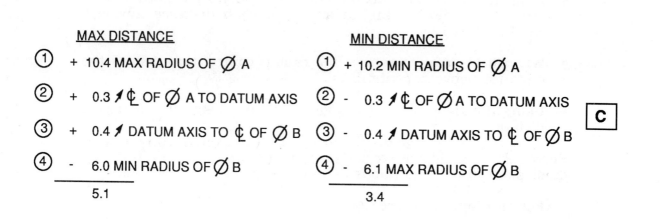

B

MAX DISTANCE

① + 10.4 MAX RADIUS OF ⌀ A

② + 0.3 ↗ ₵ OF ⌀ A TO DATUM AXIS

③ + 0.4 ↗ DATUM AXIS TO ₵ OF ⌀ B

④ − 6.0 MIN RADIUS OF ⌀ B

—————

5.1

MIN DISTANCE

① + 10.2 MIN RADIUS OF ⌀ A

② − 0.3 ↗ ₵ OF ⌀ A TO DATUM AXIS

③ − 0.4 ↗ DATUM AXIS TO ₵ OF ⌀ B

④ − 6.1 MAX RADIUS OF ⌀ B

—————

3.4

C

FIGURE 6-12 CALCULATING THE DISTANCE BETWEEN TWO PART
SURFACES CONTROLLED WITH RUNOUT

SUMMARY

A summarization of runout control information is shown in Figure 6-13.

SYMBOL	DATUM REFERENCE REQUIRED	CAN BE APPLIED TO A		CAN AFFECT VIRTUAL CONDITION	CAN USE Ⓜ MODIFIER	CAN USE Ⓢ MODIFIER	CAN OVERRIDE RULE #1
		FEATURE	FEATURE OF SIZE				
↗	YES	YES	YES	YES	NO	YES*	YES
↗↗	YES	YES	YES	YES	NO	YES*	YES

* IS AUTOMATIC PER RULE #1

FIGURE 6-13 SUMMARIZATION OF RUNOUT CONTROLS

VOCABULARY WORDS

Composite Tolerance
Runout
Circular Runout
Total Runout

THOUGHT QUESTIONS

1. Discuss the merits of specifying runout to control the location of a radius.

2. Discuss the control achieved by specifying circular runout to a surface which is perpendicular to the datum axis.

PROBLEMS AND QUESTIONS

1. List three ways of establishing a datum axis for a runout control.

   a. _____

   b. _____

   c. _____

2. When checking runout controls, the indicator is always held perpendicular to the datum axis. Circle and explain your answer.

   true _____

   false _____

3. Circular runout is a composite control. List two characteristics that circular runout affects.

   a. _____

   b. _____

4. List two differences between circular runout and total runout.

   a. _____

   b. _____

5. When checking runout controls with a dial indicator, readings are made while the part is rotated about the axis of the feature being checked. Circle and explain your answer.

   true _____

   false _____

6. True or False

T	F	Runout can use a bonus tolerance.
T	F	Runout is verified with a moveable gage.
T	F	Circular runout is a more stringent control than concentricity.
T	F	Runout can never use a shift tolerance.
T	F	Runout always requires a datum reference.

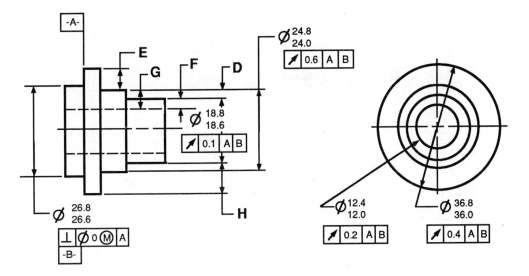

QUESTIONS 7-11   REFER TO THE ABOVE FIGURE

7. Calculate the missing information in the chart below.

DISTANCE	MAXIMUM	MINIMUM
D		
E		
F		
G		
H		

8. What would be the effect if the datum sequence was reversed in the runout callouts? _____

_____

9. If datum B was at LMC, would a shift tolerance be available?

_____

10. If the runout callouts were revised to total runout, what would be the effect on the maximum and minimum distances in the chart above? _____

11. What is the largest diameter pin that would always fit into the hole in the part above. _____

ACROSS

1. TWO CHARACTERISTICS WHICH RUNOUT CONTROLS
6. CONCENTRATIVE
7. TWO DIAMETERS SPACED _____
11. EXPERT
13. FASTENER
15. TOTAL RUNOUT TOLERANCE ZONE (2 WDS.)
16. A SMALL WOODEN PIN USED TO FASTEN THINGS
17. TYPE OF RUNOUT
18. FLOP
19. TYPE OF BRAKE
21. TOTAL INDICATOR READING (ABBR.)
25. NUMBER OF WAYS TO ESTABLISH A DATUM AXIS
27. TECHNICAL BOOK AUTHOR
28. BASKET OR BASE SUFFIX
29. SOMETHING PRECIOUS
31. CRAVE
32. RUNOUT TOLERANCE ZONE
38. BORDERED
40. NEGLECT
42. DANCE LIKE BOJANGLES
43. FEATURE RUNOUT CAN CONTROL
44. SYMBOL FOR RUNOUT

DOWN

1. _____ PRIMARY, DIAMETER SECONDARY
2. RULE APPLIES TO RUNOUT
3. TELL TALES
4. TO TALK OF SECRET MATTERS
5. FRUIT COVERING
6. GROUP
8. TYPE OF BEAR
9. COAXIAL DIAMETER CONTROL
10. BASIS FOR RUNOUT DATUM SELECTION (2 WDS.)
12. RUNOUT IS A _____ CONTROL
14. STRIPED CAT
17. EACH DIAMETER HAS ONE
19. EVERY RUNOUT CONTROL NEEDS ONE
20. _____ OFFSET
22. BABY'S BED
23. FULL INDICATOR MOVEMENT (ABBR.)
24. _____ RUNOUT
26. ANYTIME
30. _____ DIAMETER
31. DISREGARD
33. CONCEALED GUNMAN
34. PORK CUT
35. LITTLE ONE
36. SLOGAN
37. AUCTION ACTIONS
38. IMPORTANT TIME IN HISTORY
39. FALLOW _____
41. UNIT OF LENGTH

CHAPTER **7**

PROFILE CONTROLS

If your margin of safety is too great or too small, your competition will beat you.

INTRODUCTION

This chapter provides a basic understanding of profile controls. A lack of understanding of profile controls has resulted in their limited use on engineering drawings today. By studying the concepts in this chapter, you will become proficient in the interpretation and application of profile controls.

GENERAL INFORMATION

This chapter discusses two types of profile controls, profile of a line and profile of a surface. The symbol for each is shown in Figure 7-1. Profile controls can be used to limit the form, size, or orientation of a part feature. There are examples and information unique to each type of control, but the following information applies to both types of profile controls.

The outline of an object in a given plane is referred to as its *profile*. A *true profile* is the exact profile of a geometric shape as described by basic dimensions. A *profile tolerance* specifies a uniform boundary along the true profile within which all the elements of the considered surface element or elements must lie. A profile tolerance can be applied to all surface elements simultaneously (as in profile of a surface), or to individual surface elements (as in profile of a line) taken at various cross sections through the part.

Profile controls are unique because they are the only geometric tolerance which can be used as a form control (without a datum) or as a related feature tolerance (with a datum).

REMEMBER

Profile can be used as a form control or a related feature tolerance.

ADVANTAGES OF PROFILE CONTROLS

Three advantages from using profile controls are:

- Clear definition of the tolerance zone

- Communicates datums and datum sequence

- Eliminates accumulation of tolerances

PROFILE OF A SURFACE

PROFILE OF A LINE

FIGURE 7-1 PROFILE CONTROL SYMBOLS

A profile control can be applied to any type of part feature (i.e. surface, irregular shape, cylinder, etc.), but the true profile of the feature must be defined with basic dimensions.

A unique aspect of profile controls is that the tolerance zone can be specified to apply either unilaterally or bilaterally. Figure 7-2 illustrates this concept.

REMEMBER

Whenever a profile tolerance is specified, the true profile of the feature must be defined with basic dimensions.

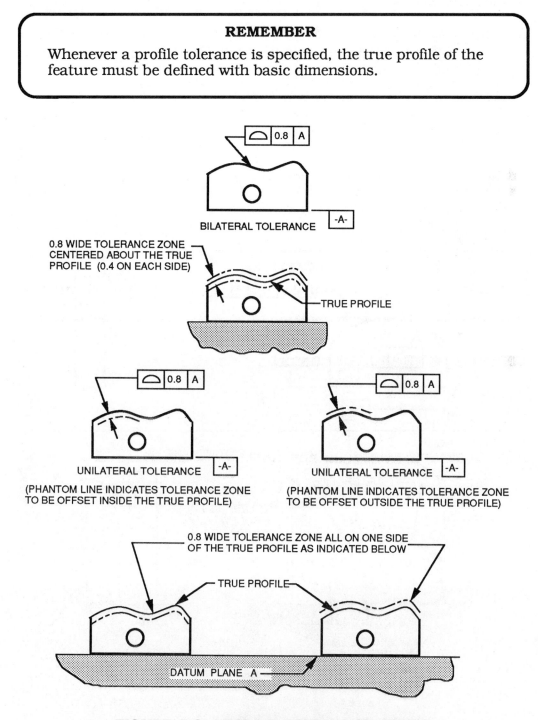

FIGURE 7-2 PROFILE TOLERANCE ZONES

PROFILE OF A SURFACE

When a profile of a surface control is specified, the tolerance zone is three dimensional. It extends along the entire length, width, and depth of the toleranced feature simultaneously. The tolerance zone is two parallel boundaries, offset from the true profile by a specified amount. Figure 7-3 illustrates the above concept.

> **REMEMBER**
>
> A profile of a surface control limits all surface elements simultaneously.

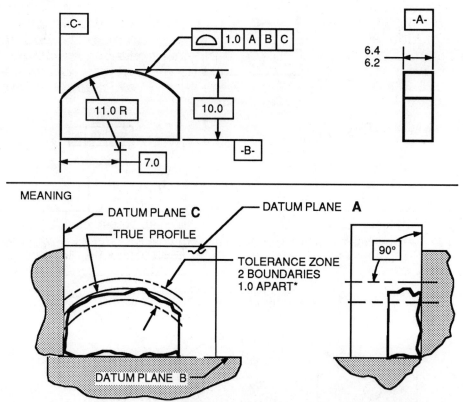

FIGURE 7-3 PROFILE OF A SURFACE

SUMMARY OF INFORMATION FROM FIGURE 7-3

- The profile (of a surface) control applies RFS (per Rule #3).

- Bonus tolerance concepts are not applicable.

- The shape of the tolerance zone is two parallel boundaries offset an equal distance (above and below) from the true profile.

- The profile control limits the size of the part as well as the orientation and form of the toleranced feature.

APPLICATIONS

Common applications for profile of a surface controls include: defining boundaries for polygons, irregular shapes (mixture of arcs and lines), coplanar surfaces, and relating a dimensional measurement (location of a feature) to a datum reference frame. The following paragraphs give examples and information about each type of control.

Profile of a surface is often used to establish acceptable boundaries for polygonal shapes. When defining the parameters for a polygonal shaped part feature defined with coordinate dimensions, an accumulation of tolerances often becomes a problem for both part function and inspection. Using profile of a surface to define the polygon boundaries can eliminate many of the problems incurred with coordinate dimensioning. Figure 7-4 illustrates an example of profile dimensioning of a polygonal shape.

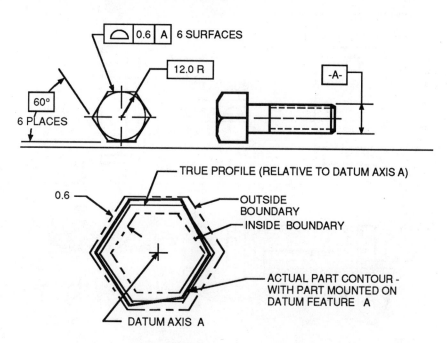

FIGURE 7-4 PROFILE OF A SURFACE APPLICATION

SUMMARY OF INFORMATION FROM FIGURE 7-4

- Basic dimensions are used to define the true profile of the polygon.

- The profile control applies RFS (per rule #3).

- The shape of the tolerance zone is two parallel boundaries offset an equal distance (above and below) from the true profile.

- This dimensioning method eliminates tolerance accumulation.

- The order of addressing the part to the datum reference frame for checking is specified in the feature control frame.

A profile of a surface control is often used to define the acceptable limits for coplanar surfaces. *Coplanar surfaces* are two (or more) surfaces having all elements in the same plane. When profile of a surface is applied to coplanar surfaces it establishes a set of parallel planes within which all of the elements of both surfaces must lie. Profile of a surface applied to coplanar surfaces serves the same purpose as flatness applied to a single surface. Whenever profile of a surface is applied to coplanar surfaces the tolerance zone is unilateral.

When applying profile to coplanar surfaces, if more than two surfaces are being controlled, a datum reference should be used. The surfaces which are intended to contact the datum plane are designated as datum surfaces. The part is located on these surfaces when verifying the profile of the coplanar surfaces. Figure 7-5 illustrates the above concepts.

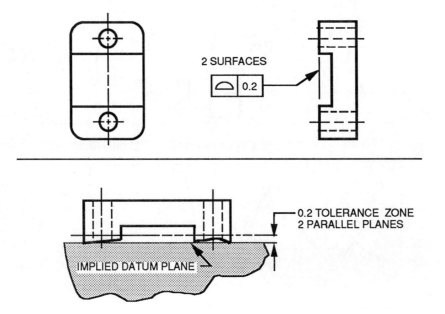

FIGURE 7-5 PROFILE OF A SURFACE APPLICATION

SUMMARY OF INFORMATION FROM FIGURE 7-5

- When checking the coplanarity of the surfaces, both surfaces must be in contact with the datum plane.

- The tolerance zone is unilateral.

- The profile tolerance applies RFS (per Rule #3).

- The flatness of each surface is limited to within the profile tolerance zone.

Figure 7-6 shows an example of using profile tolerancing to relate a dimensional measurement to a datum reference frame. This method of dimensioning should be considered whenever it is desirable to relate dimensional measurements to a datum reference frame. It allows the designer to communicate a sequence for addressing the part to the datums for checking specified dimensions.

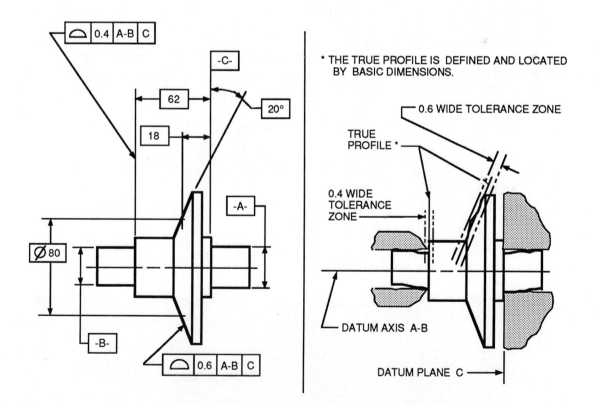

FIGURE 7-6 PROFILE OF A SURFACE APPLICATION

SUMMARY OF INFORMATION FROM FIGURE 7-6

-   The profile tolerance applies RFS (per Rule #3).

-   The sequence for bringing the part to the datum reference frame for making measurements is specified in the profile callout.

LEGAL SPECIFICATION TEST

For a profile of a surface control to be a legal specification, it must satisfy the following conditions:

•   Basic dimensions must be used to define the true profile of the toleranced feature.

•   The profile control must be applied to the true profile of the toleranced feature.

•   No modifiers may be used in the tolerance portion of the feature control frame.

•   If datum references are used, basic dimensions must relate the true profile to the datums specified.

A simple test for verifying if a profile of a surface control is a legal specification is shown in Figure 7-7.

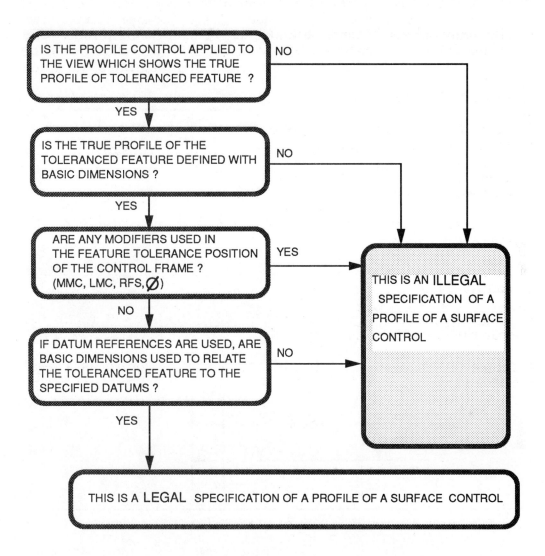

FIGURE 7-7 TEST FOR LEGAL SPECIFICATION -
PROFILE OF A SURFACE

PROFILE OF A LINE

When a profile of a line control is specified, it limits the boundaries of individual line elements of a surface. The tolerance zone is two dimensional. It extends for the entire length of the true profile. The tolerance zone is two parallel straight or curved lines, offset from the true profile by the specified amount. Figure 7-8 illustrates the above concept.

REMEMBER

Profile of a line applies to each line element of the considered feature individually.

DRAWING

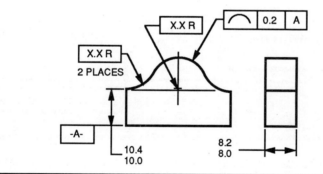

MEANING

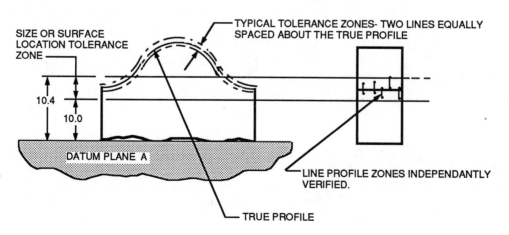

FIGURE 7-8 PROFILE OF A LINE AND SIZE CONTROL

SUMMARY OF INFORMATION FROM FIGURE 7-8

- Basic dimensions are used to define the true profile.

- The profile control applies RFS (per Rule #3).

- The shape of the tolerance zone is two parallel boundaries (line elements) offset an equal distance above and below the true profile.

- The profile control limits the orientation and form of the toleranced feature.

LEGAL SPECIFICATION TEST

For a profile of a line control to be a legal specification, it must satisfy the following conditions:

- Basic dimensions must be used to define the true profile of the toleranced feature.

- The profile control must be applied to the true profile of the toleranced feature.

- No modifiers may be used in the tolerance portion of the feature control frame.

- If toleranced dimensions are used to locate the considered feature, the profile tolerance value must be a refinement of the tolerance zone established by the toleranced dimensions.

- If the profile control is used as a form control (no datum reference), the tolerance value must be a refinement of the size tolerance of the feature.

A simple test for verifying if a profile of a line control is a legal specification is shown in Figure 7-9.

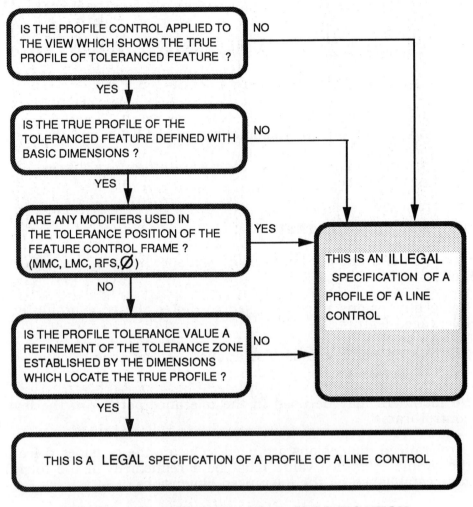

FIGURE 7-9   TEST FOR LEGAL SPECIFICATION -
PROFILE OF A LINE

SUMMARY

A summarization of profile concepts is shown in Figure 7-10.

SYMBOL	DATUM REFERENCE REQ'D OR PROPER	CAN BE APPLIED TO A		CAN USE MODIFIER	CAN USE MODIFIER	CAN OVERRIDE RULE #1	CAN USE BONUS TOLERANCE CONCEPTS	CAN USE SHIFT TOLERANCE CONCEPTS
		FEATURE	FEATURE OF SIZE					
⌒	YES *	YES	NO	NO	YES **	NO	NO	NO
⌒	YES *	YES	NO	NO	YES **	NO	NO	NO

*   CAN BE USED WITH OR WITHOUT A DATUM REFERENCE
** IS AUTOMATIC PER RULE #3

FIGURE 7-10   SUMMARIZATION OF PROFILE CONTROLS

VOCABULARY WORDS

> True Profile
> Profile Tolerance
> Unilateral Tolerance Zone
> Bilateral Tolerance Zone
> Coplanar Surfaces

THOUGHT QUESTIONS

1. Discuss the implications of allowing the use of the MMC modifier in the datum reference portion of a profile control callout.

2. Discuss the similarities of a profile of a line control and a straightness of a line element control.

PROBLEMS AND QUESTIONS

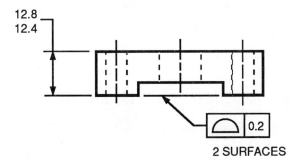

2 SURFACES

QUESTIONS 1-6 REFER TO THE ABOVE FIGURE

1. Describe the tolerance zone for the profile callout _____

_____

2. What is the virtual condition of the height of the part?

_____

3. Is the tolerance zone for the profile callout bilateral, or unilateral?

_____

4. What is the flatness of the surfaces on the bottom of the part limited to? _____

189

5. Is Rule #1 overridden by the profile tolerance?
   _____

6. Can profile of a surface be applied to a feature of size? _____

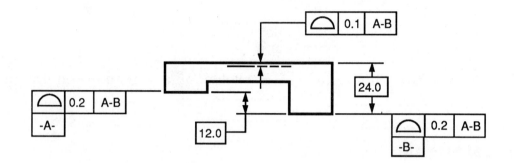

QUESTIONS 7-12 REFER TO THE ABOVE FIGURE

7. What is the virtual condition for the height of the part?
   _____

8. Are the profile tolerance zones for surfaces A & B unilateral or bilateral? _____

9. If the 12.0 basic dimension was replaced with a 12.0-12.2 toleranced dimension, how would the profile tolerance zones be affected? _____

10. If the 24.0 basic dimension was replaced with a 23.9-24.0 dimension, how would the boundaries for the height of the part be affected? _____

11. Which datum is primary A or B? _____

12. Draw and label a cartoon gage for verifying the profile tolerances.

190

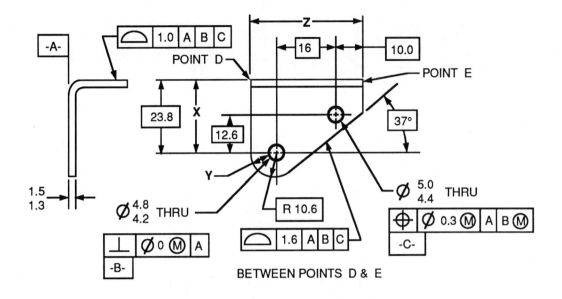

QUESTIONS 13-20   REFER TO THE ABOVE FIGURE

13.   What is the maximum distance X can be?   _____

14.   What is the minimum distance Y can be?   _____

15.   What is the maximum distance Y can be?   _____

16.   What is the minimum distance X can be?   _____

17.   What is the maximum distance Z can be?   _____

18.   What is the minimum distance Z can be?   _____

19.   What controls the flatness of datum feature A?   _____

20.   Draw and label a cartoon gage for checking the profile callouts.

191

ACROSS

1. TOWEL ID
3. IN ADDITION
6. ATTACH
8. SPEEDY
10. _____ DIMENSION
12. RULE APPLIES TO PROFILE CONTROLS
17. ANCIENT
18. TOWARD
19. OUTLINE DESCRIBED WITH
    BASIC DIMENSIONS (2 WDS.)
21. CLOSE
22. LASTING INJURY
26. MILD
27. PROFILE _____
29. SWEETHEART
31. SMALL
33. UNILATERAL OR _____ ZONE
34. USE LIKE A KNIFE
35. DANGLE
36. NOISY
38. CHARACTERISTIC WHICH
    PROFILE CONTROLS
39. "THE _____ STRIKES BACK"
40. DEFAULT FOR PROFILE CONTROLS (INT)
43. PROFILE CAN CONTROL
    _____ SURFACES
44. NUCLEUS
46. OUTLINE OF AN OBJECT
47. WESTERN SHOW
48. INTELLIGENT

DOWN

2. STEP
4. ENDED
5. INCREASE
7. EVALUATE
9. PEEP OUT
11. TOLERANCE _____
13. _____ ELEMENT
14. _____ FEATURE CONTROL
15. FORE AND _____
16. NOT THE TRUTH
18. GROUP OF DRAFT ANIMALS
20. USED FOR COOKING
22. VIRGINIA, E.G.
    23. METHOD OF TRANSPORTATION
24. A COLOR
25. TECHNICAL BOOK AUTHOR
27. OHIO CITY
28. PROFILE TOLERANCE ZONE
30. USE WITH PROFILE CONTROLS
32. LESSER
34. PROFILE CAN BE APPLIED TO A _____
37. STAMPING TOOLS
40. RENTED AT A HOTEL
41. STUPEFY
42. ANY MONKEY
45. VERTEBRAE ORGAN OF HEARING

192

CHAPTER 8

GLOSSARY

Learn from the mistakes of others. You can never live long enough to make them all yourself.

INTRODUCTION

The information in this chapter is valuable as reference material.

GLOSSARY

3-2-1 RULE - The 3-2-1 Rule defines the minimum number of points of contact required for a primary, secondary, and tertiary datum feature with their respective datum planes.

ANGULARITY TOLERANCE - An Angularity Tolerance is the amount which a surface, centerplane, or axis is permitted to vary from its specified exact angle.

BASIC DIMENSION - A Basic Dimension is a numerical value used to describe theoretically exact characteristics of a feature or datum target.

BILATERAL TOLERANCE ZONE - A tolerance zone which allows variation in both directions from the specified dimension.

BONUS TOLERANCE - When the actual feature-of-size departs from MMC, an increase in the stated tolerance, equal to the amount of the departure, is permitted. This increase or extra tolerance is called the Bonus Tolerance.

CARTOON GAGE - A Cartoon Gage is a sketch of a functional gage.

CIRCULAR RUNOUT - Circular Runout is a composite control affecting both the form and location of circular elements of a part feature.

CIRCULARITY TOLERANCE - A Circularity Tolerance is the amount which surface elements of the diameter may vary from a theoretical circle.

CO-DATUMS - When two datum features of equal importance are used to establish a single datum plane or axis, they are called Co-Datums.

COAXIAL FEATURES-OF-SIZE - Coaxial Features-Of-Size are two (or more) features-of-size, normally diameters, whose axes are coincident.

COMPOSITE CONTROL - A composite control is used to control cumulative variations of a part feature. Circular and total runout are composite controls.

CONCENTRICITY - The condition where the axes of all cross-sectional elements of a feature-of-size, and surface of revolution are common to the axis of a datum feature.

COPLANAR SURFACES - Coplanar Surfaces are two (or more) surfaces having all elements in the same plane.

CYLINDRICITY TOLERANCE - A Cylindricity Tolerance is the amount which surface elements of a cylinder may be allowed to vary from a theoretically perfect cylinder.

DATUM - A Datum is theoretically exact point, line, axis or plane which indicates the orgin of a specified dimensional relationship between a toleranced feature and a designated feature on a part.

DATUM AXIS - Where a feature-of-size is specified as a datum feature, the surface or surfaces of that feature-of-size are used to establish a Datum Axis or Centerplane.

DATUM FEATURE - A Datum Feature is an actual part feature which contacts, or is used to establish, a datum.

DATUM PLANE - A Datum Plane is a theoretical plane which contact three high points of datum feature.

DATUM REFERENCE FRAME* - A Datum Reference Frame* is a set of three mutually perpendicular planes.

DATUM SHIFT - The looseness of movement, between the part datum feature and the gage, is called Datum Shift.

DATUM TARGETS - Datum Targets are designated points, lines or areas of contact used to locate a part in a datum reference frame.*

FEATURE - A Feature is general term applied to a physical portion of a part, such as a surface, hole, or slot.

FEATURE CONTROL FRAME - A Feature Control Frame is a rectangle which is divided into compartments within which the geometric characteristic symbol, tolerance value, modifiers, and datum references are placed.

FEATURE-OF-SIZE - A Feature-of-Size is one cylindrical or spherical surface or a set of parallel surfaces, each of which is associated with a size dimension.

FIXED GAGE - A fixed Gage is a gage that has no moving parts.

FLATNESS TOLERANCE - A Flatness Tolerance is the amount which surface elements are permitted to vary from a theoretical plane.

FUNCTIONAL ANALYSIS - A Functional Analysis is a process where a designer identifies the functions of a part and uses this information to establish the actual part dimensions and tolerances.

FUNCTIONAL DIMENSIONING - Functional Dimensioning is a philosophy of dimensioning and tolerancing a part based on how it functions.

FUNCTIONAL GAGE - A Functional Gage is a gage which verifies the functional requirements of part features.

GEOMETRIC DIMENSIONING AND TOLERANCING (G.D.&T.) - Geometric Dimensioning and Tolerancing (G.D.&T.) is a dual purpose system. First it is a set of standard symbols which are used to define part features and their tolerance zones. The symbols and their interpretation are documented by the ANSI Y14.5M-1982 standard. Second, and of equal importance, Geometric Dimensioning and Tolerancing is a philosophy of defining a part based on how it functions.

LEAST MATERIAL CONDITION - When a feature-of-size contains the minimum amount of material, it is in its Least Material Condition (LMC).

LOCATION DIMENSION - A Location Dimension is a dimension which locates the centerline or centerplane of a part feature relative to another part feature, centerline, or centerplane.

MAXIMUM MATERIAL CONDITION - When a feature-of-size contains the most amount of material, it is in its Maximum Material Condition (MMC).

ORIENTATION CONTROL - Orientation controls establish the orientation of features to one another.

PARALLELISM TOLERANCE - A Parallelism Tolerance is the amount which a surface, centerplane, or axis is permitted to vary from the parallel state.

PERPENDICULARITY TOLERANCE - A Perpendicularity Tolerance is the amount which a surface, or axis, or centerplane is permitted to vary from being perpendicular.

PROFILE - The outline of an object in a given plane is referred to as its Profile.

PROFILE TOLERANCE - A Profile Tolerance specifies a uniform boundary along the true profile within which all the elements of the considered surface elements of elements must lie.

REGARDLESS OF FEATURE SIZE - Regardless Of Feature Size (RFS) is when a geometric tolerance (or datum) applies independent of the feature size.

RULE #1 (LIMITS OF SIZE RULE) - For features-of-size, when only a tolerance of size is specified, the surfaces shall not extend beyond a boundary (envelope) or PERFECT FORM AT MMC.

RULE #2 (TOLERANCE OF POSITION RULE) - For tolerances of position, S, L, or M must be specified in the feature control frame with respect to the tolerance value, datum reference, or both, as applicable.

RULE #3 - For other than a tolerance of position, RFS applies with respect to the tolerance, datum reference, or both, where no modifier is specified. MMC must be specified in the feature control frame when it is appropriate and desired.

RULE #5 - Instances where the virtual condition is used to determine the gage size for an MMC datum are referred to as Rule #5 in the 1973 edition of ANSI Y14.5.

RUNOUT - Runout is a composite control affecting both the form and location of a part feature relative to a datum axis.

STRAIGHTNESS TOLERANCE - A Straightness Tolerance is the amount a surface line element of a feature is permitted to vary from a theoretically straight line.

TOLERANCE OF - POSITION - The total permissible variation in the location of a feature-of-size about its true position.

TOTAL RUNOUT - Total Runout is a composite control affecting the form and location of all surface elements simultaneously.

TRUE POSITION - A term used to describe the exact (perfect) location of a point, line, or plane (normally the center) of a feature-of-size in relationship with a datum reference frame and/or other features-of-size. Basic dimensions are used to establish the true position of features-of-size on drawings.

TRUE PROFILE - A True Profile is the exact geometric shape of a profile as described by basic dimensions.

UNILATERAL TOLERANCE ZONE - A tolerance zone which allows variation in only one direction from the specified dimension.

VIRTUAL CONDITION - Virtual Condition is the theoretical extreme boundary of a feature-of-size generated by the collective effects of MMC and any applicable geometric tolerances.

CHAPTER **9**

CLASS EXERCISES

Most of us learn more from our own mistakes than we do from good examples set by others.

INTRODUCTION

These exercises are to be done as a class activity. Do not work ahead. The exercises contain a few intentional errors. They are designed to create class discussions   Remember, it's not who is right, but what is right that counts on these exercises.

TAKE TIME TO THINK

IT IS THE SOURCE OF ALL POWER

EXERCISE 1
ENGINEERING DRAWINGS

1. What is an engineering drawing?_____

_____

_____

_____

2. List 5 different job classifications that use engineering drawings in your company.

_____

_____

_____

_____

_____

3. Does fully defining a part make it more expensive? Why? _____

_____

_____

_____

_____

4. List three possible outcomes of missing dimensions on a drawing.

_____

_____

_____

_____

EXERCISE 2

FEATURE - FEATURE OF SIZE

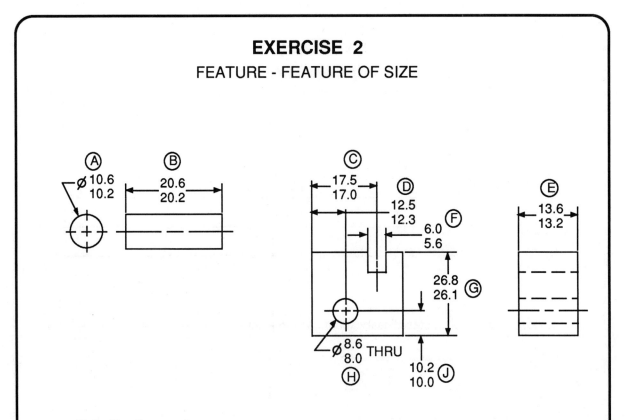

Using the figures above,
fill in the chart below.

Dimension	Feature	Feature of size	Location Dimension
A			
B			
C			
D			
E			
F			
G			
H			
J			

EXERCISE 3
MMC - LMC

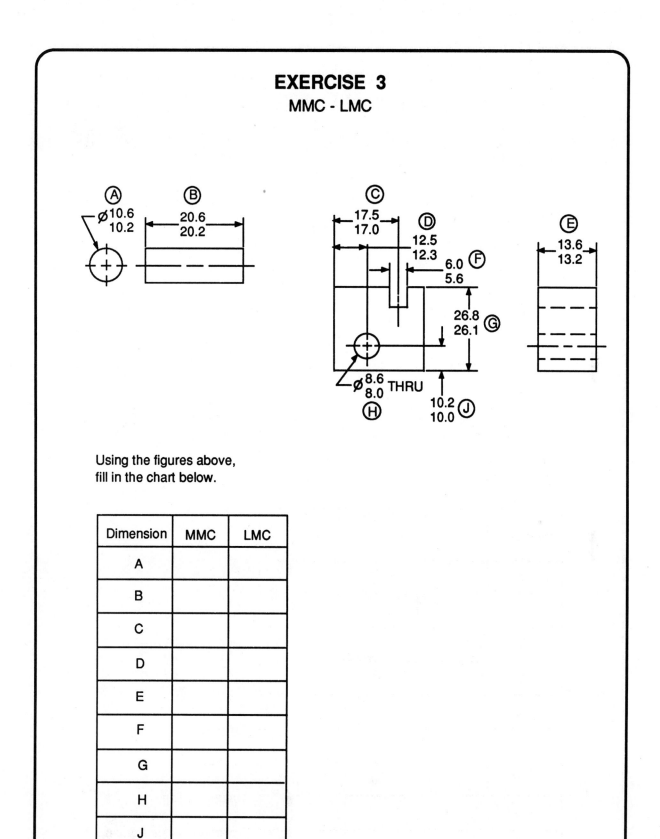

Using the figures above,
fill in the chart below.

Dimension	MMC	LMC
A		
B		
C		
D		
E		
F		
G		
H		
J		

EXERCISE 4
SYMBOLS + FEATURE CONTROL FRAME

1. ON EACH BLANK, SKETCH THE APPROPRIATE SYMBOL,
   SELECTING FROM THE SYMBOLS SHOWN.

a. _____ Flatness   h. _____ Profile of a line

b. _____ Total Runout   i. _____ Cylindricty

c. _____ Circularity   j. _____ Tolerance of position

d. _____ Straightness   k. _____ Circular Runout

e. _____ Parallelism   l. _____ Profile of a surface

f. _____ Angularity   m. _____ Concentricty

g. _____ Perpendicularity   n. _____ Regardless of feature size

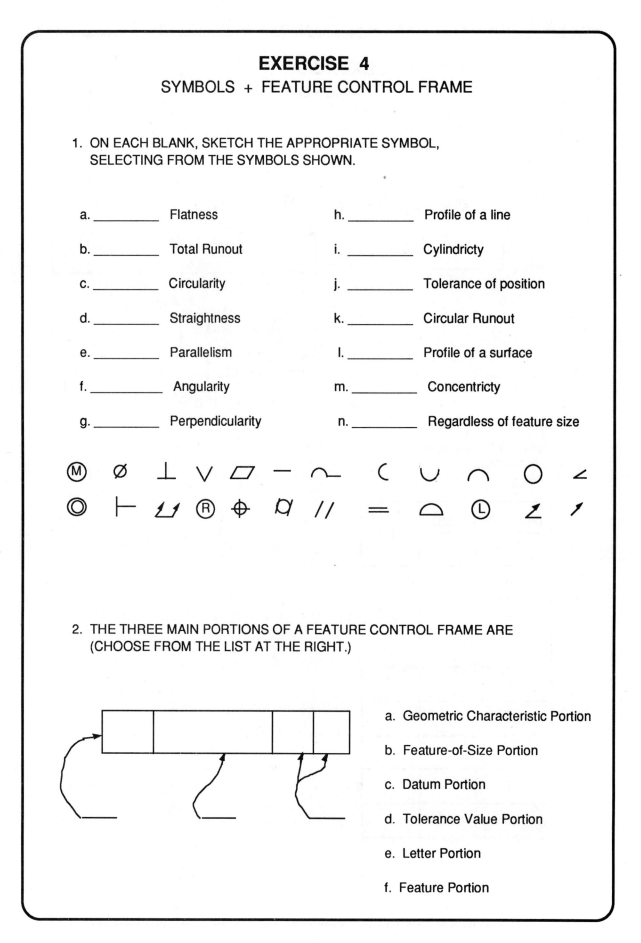

2. THE THREE MAIN PORTIONS OF A FEATURE CONTROL FRAME ARE
   (CHOOSE FROM THE LIST AT THE RIGHT.)

a. Geometric Characteristic Portion

b. Feature-of-Size Portion

c. Datum Portion

d. Tolerance Value Portion

e. Letter Portion

f. Feature Portion

EXERCISE 5
RULE #1

FILL IN THE BLANKS

1. If the ∅ of the pin is / the allowable straightness error is

If the ∅ of the pin is	the allowable straightness error is
10.4	
10.6	
10.2	
10.5	
10.7	
10.1	

2. Does rule #1 apply to the 26.0 - 26.2 dimension? _____

3. If you were manufacturing the above pin and the process being used could not maintain the straightness of the pin to less than 0.3, what diameter would you have to produce the pins to make good parts? _____

4. State Rule #1. _____

_____

_____

EXERCISE 6
RULE #1

SURFACE A

12.2
11.8

16.2
16.0

SURFACE B

C

FILL IN THE BLANKS

1. If the block was at MMC, what would

    a. its straightness be limited to? _____

    b. the flatness of surface A be limited to? _____

    c. the flatness of surface B be limited to? _____

2. If the block was at LMC, the flatness of surface A would be limited to _____

and the flatness of surface B would be limited to _____. Could both of these

flatness conditions exist simultaneously? _____ Why? _____

_____

3. If the block was at LMC and the flatness of surface A was 0.3, the flatness of

surface B would be limited to. _____.

4. If the 11.8 - 12.2 dim. and the 16.0 - 16.2 dim. were both at MMC, what would

be the tolerance on angle C? _____

_____

5. What is the minimum setting of parallel plates that the block would always pass

thru if it was produced at MMC? _____

    At LMC? _____

EXERCISE 7
BASIC DIMENSION

FILL IN THE BLANKS OF THE PARAGRAPH WITH THE BEST CHOICE FROM THE WORDS BELOW.

_____ dimensions are often designated by enclosing the dimension in a _____. There are _____ types of basic dimensions. One type describes the theoretically _____ characteristics of part features. The second type describes _____ dimensions. Part features defined with basic dimensions must have a _____ to describe the allowable _____ for the part feature. Basic dimensions which describe _____ dimensions do not require a _____ because gages are considered to have _____ tolerance.

Tolerance	Gagemakers	Two
Rectangle	Gage	Exact
Geometric tolerance	Geometric Tolerance	Bonus Tolerance
Basic	Datum	Gage

EXERCISE 8
BONUS / VIRTUAL CONDITION

-A-

Ø 10.6 / 10.4

16.2 / 16.0

⟂ Ø 0.1 Ⓜ A

B

— 0.5 Ⓜ

FILL IN THE BLANKS

1. How much bonus tolerance is permissible with the callout at location B? _____

2. How much bonus tolerance is permissible with the $\frac{16.2}{16.0}$ dim? _____

3. If the actual hole dia. on a part was 10.5, how much bonus tolerance would be

   available? _____

4. If the actual width of a part was 16.0, how much bonus tolerance would be

   available? _____

5. The virtual condition of the 16.0 - 16.2 dimension is _____.

6. The virtual condition of the 10.6 - 10.4 diameter hole is _____.

EXERCISE 9
FLATNESS

SURFACE A

[⟋ | 0.2]

26.5
26.1

SURFACE B

FILL IN THE BLANKS

1. Describe the tolerance zone for flatness. _____

_____

2. How is the flatness tolerance zone located with respect to the part? _____

_____

3. Can a flatness symbol override rule #1? _____

4. What is the maximum possible flatness error of surface A? _____

   Surface B? _____

   Can both of these flatness errors exist simultaneously, why? _____

   _____

5. What is the virtual condition of the height of the block? _____

6. If the flatness symbol is revised to [⟋ | 0.2 Ⓜ] , what is the effect on the part?

   _____

   _____

7. If the flatness symbol is revised to [⟋ | 0.4] , what is the effect on the part?

   _____

   _____

EXERCISE 10
STRAIGHTNESS - APPLIED TO A FEATURE

SURFACE **B**

| — | 0.1 |

22.2
22.0

SIDE

SURFACE **A**

FILL IN THE BLANKS

1. The straightness symbol is applied to a _____
   feature/feature of size

2. Does the straightness control override rule #1? _____

3. Describe the tolerance zone for the straightness callout. _____

   _____

   _____

4. What is the virtual condition of the height of the block? _____

5. Can the straightness be checked with a fixed gage? _____

   Why? _____

6. What is the flatness of surface **B** limited to? _____

7. What would be the effect on the part if the straightness control was

   revised to | — | 0.1 Ⓜ | ? _____

   _____

EXERCISE 11
STRAIGHTNESS - APPLIED TO A FEATURE-OF-SIZE

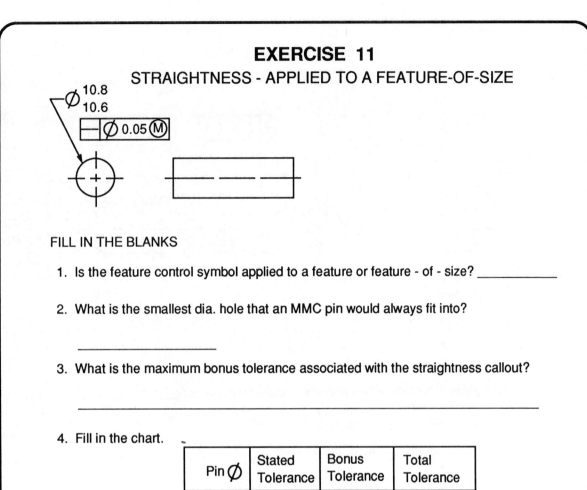

FILL IN THE BLANKS

1. Is the feature control symbol applied to a feature or feature - of - size? _____

2. What is the smallest dia. hole that an MMC pin would always fit into?

   _____

3. What is the maximum bonus tolerance associated with the straightness callout?

   _____

4. Fill in the chart.

Pin $\emptyset$	Stated Tolerance	Bonus Tolerance	Total Tolerance
10.7			
10.6			
10.5			
10.8			
10.9			

5. If the straightness modifier is removed, what will be the bonus tolerance

   associated with the dia. of the pin? _____

6. What is the virtual condition of the pin $\emptyset$?_____

7. Describe the gage for checking the straightness of the pin.

   _____

211

EXERCISE 12
CIRCULARITY

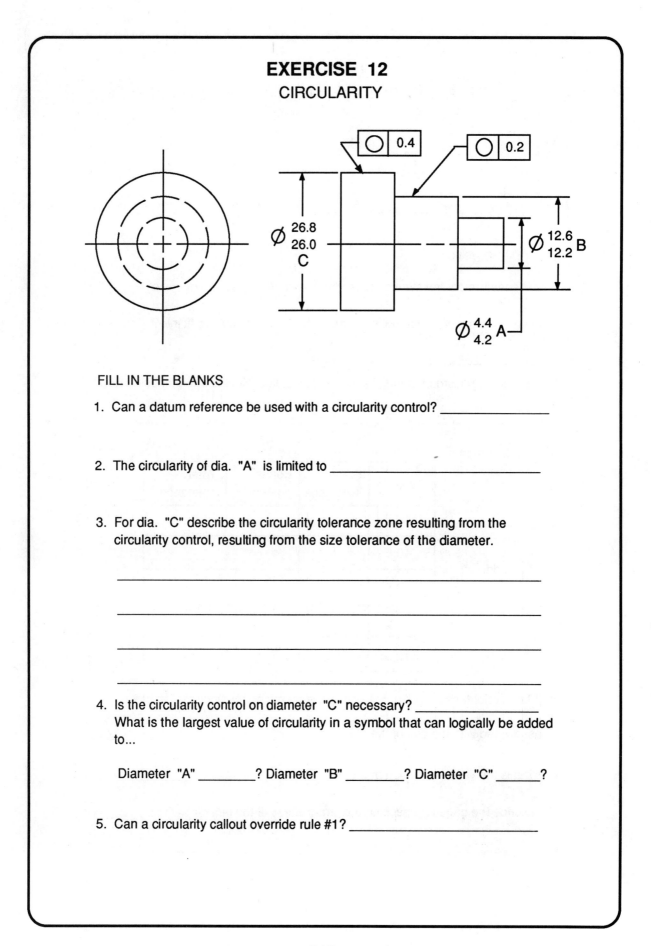

FILL IN THE BLANKS

1. Can a datum reference be used with a circularity control? _____

2. The circularity of dia. "A" is limited to _____

3. For dia. "C" describe the circularity tolerance zone resulting from the circularity control, resulting from the size tolerance of the diameter.

   _____

   _____

   _____

   _____

4. Is the circularity control on diameter "C" necessary? _____
   What is the largest value of circularity in a symbol that can logically be added to...

   Diameter "A" _____? Diameter "B" _____? Diameter "C" _____?

5. Can a circularity callout override rule #1? _____

EXERCISE 13
CYLINDRICITY

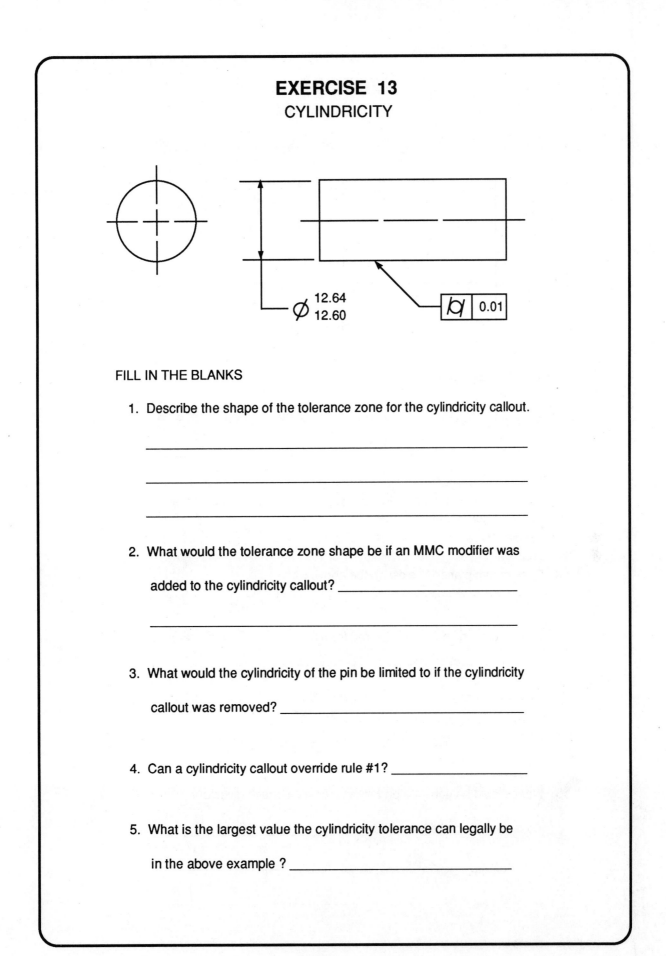

Ø 12.64
12.60

⌀ 0.01

FILL IN THE BLANKS

1. Describe the shape of the tolerance zone for the cylindricity callout.

_____

_____

_____

2. What would the tolerance zone shape be if an MMC modifier was

   added to the cylindricity callout? _____

   _____

3. What would the cylindricity of the pin be limited to if the cylindricity

   callout was removed? _____

4. Can a cylindricity callout override rule #1? _____

5. What is the largest value the cylindricity tolerance can legally be

   in the above example ? _____

EXERCISE 14
PLANAR DATUMS

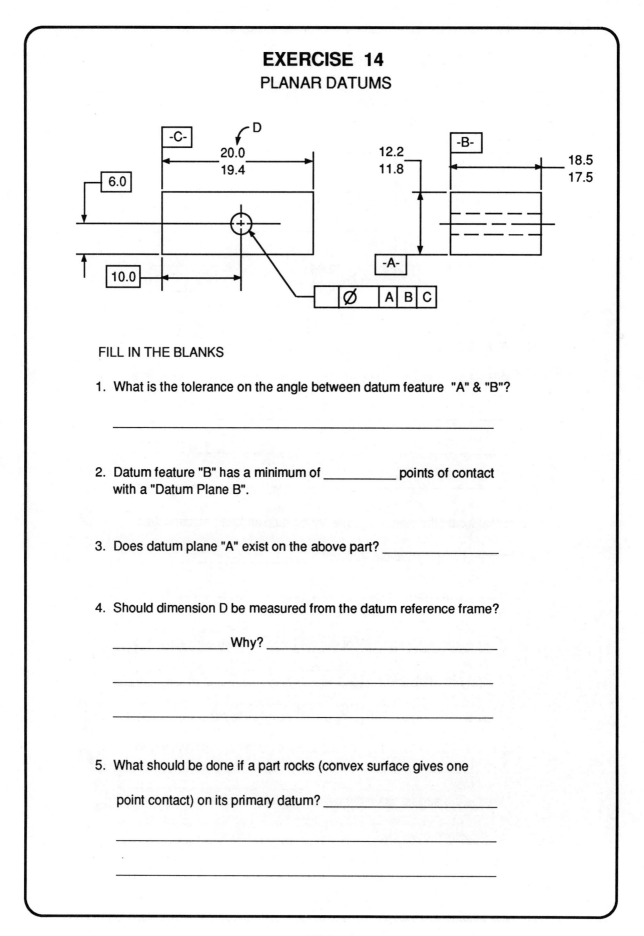

FILL IN THE BLANKS

1. What is the tolerance on the angle between datum feature "A" & "B"?

   _____

2. Datum feature "B" has a minimum of _____ points of contact
   with a "Datum Plane B".

3. Does datum plane "A" exist on the above part? _____

4. Should dimension D be measured from the datum reference frame?

   _____ Why? _____

   _____

   _____

5. What should be done if a part rocks (convex surface gives one

   point contact) on its primary datum? _____

   _____

   _____

EXERCISE 15
DATUM TARGETS

FILL IN THE BLANKS

   1. What is the tolerance on a basic dimension? _____

   2. Do basic dimensions exist on the part? _____

        _____

   3. How does a basic dimension differ from a gage dimension? _____

        _____

        _____

   4. When should a datum target be used? _____

        _____

        _____

   5. Three types of datum targets are...

                    _____

                    _____

                    _____

   6. Would it be proper to use four datum target points to establish a primary

      datum plane? _____ Why? _____

        _____

        _____

EXERCISE 16
DATUMS RFS

IN EACH CASE DESCRIBE HOW TO ESTABLISH DATUM
"A " AS PRIMARY RFS

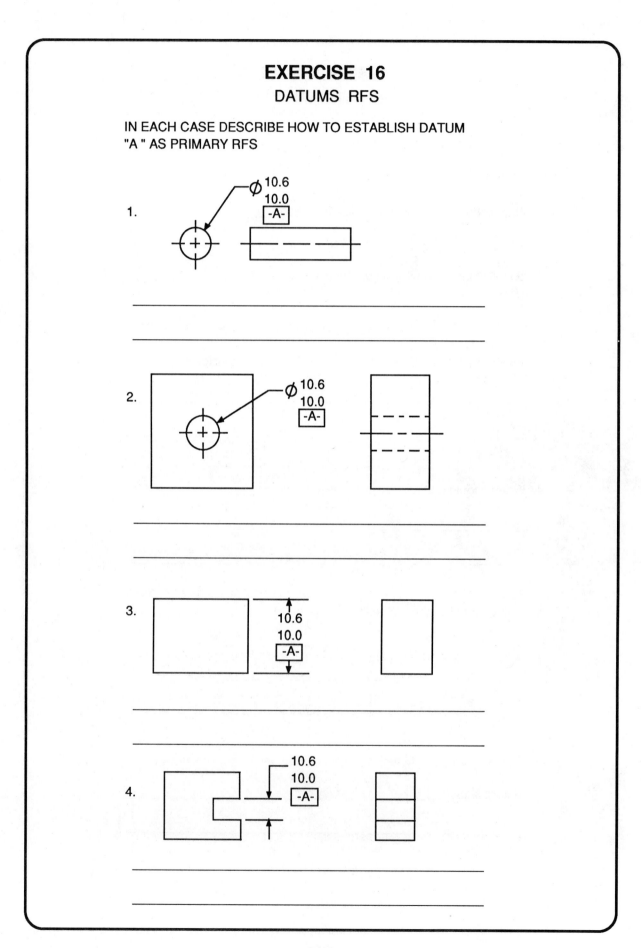

EXERCISE 17
DATUM MMC

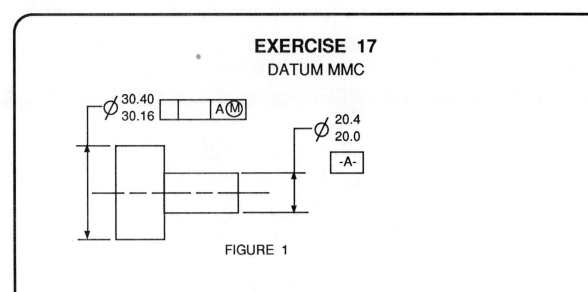

FIGURE 1

1. Draw and dimension the gage for establishing datum axis "A" for the part shown in figure 1.

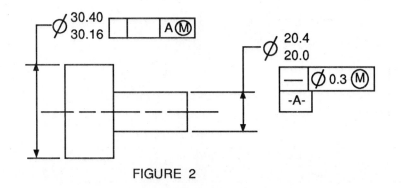

FIGURE 2

2. Draw and dimension the gage for establishing datum axis "A" for the part shown in figure 2.

3. What is the maximum amount of shift possible for the above

   parts? _____ _____
                    figure 1   figure 2

4. Whenever a datum is referenced at MMC, it can be simulated with

   a _____ gage
       fixed/variable

5. Whenever a datum is referenced at RFS, it can be simulated with

   a _____ gage
       fixed/variable

EXERCISE 18
DATUM PRECEDENCE

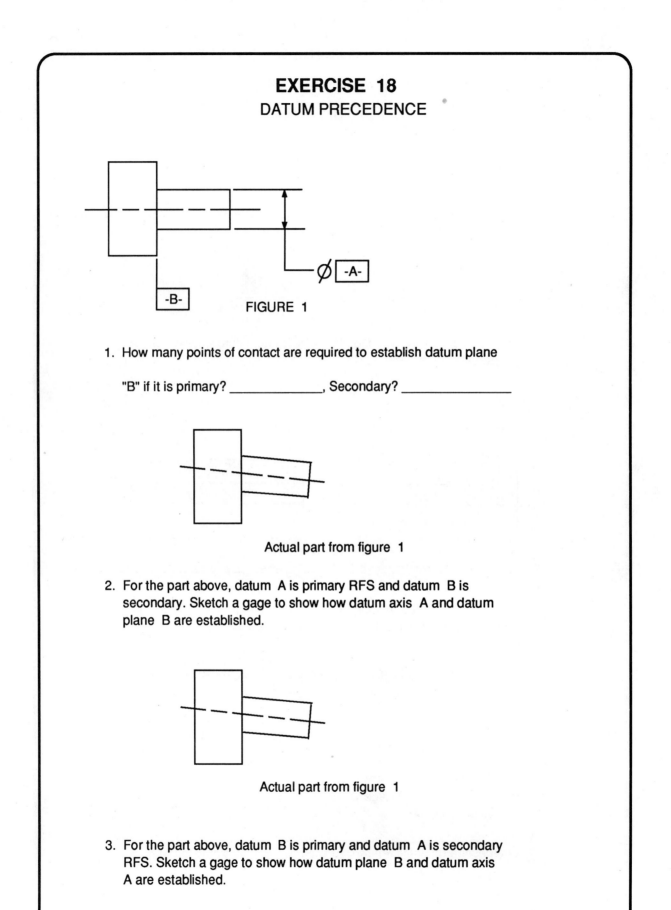

FIGURE 1

1. How many points of contact are required to establish datum plane

   "B" if it is primary? _____, Secondary? _____

Actual part from figure 1

2. For the part above, datum A is primary RFS and datum B is secondary. Sketch a gage to show how datum axis A and datum plane B are established.

Actual part from figure 1

3. For the part above, datum B is primary and datum A is secondary RFS. Sketch a gage to show how datum plane B and datum axis A are established.

EXERCISE 19
DATUM APPLICATION

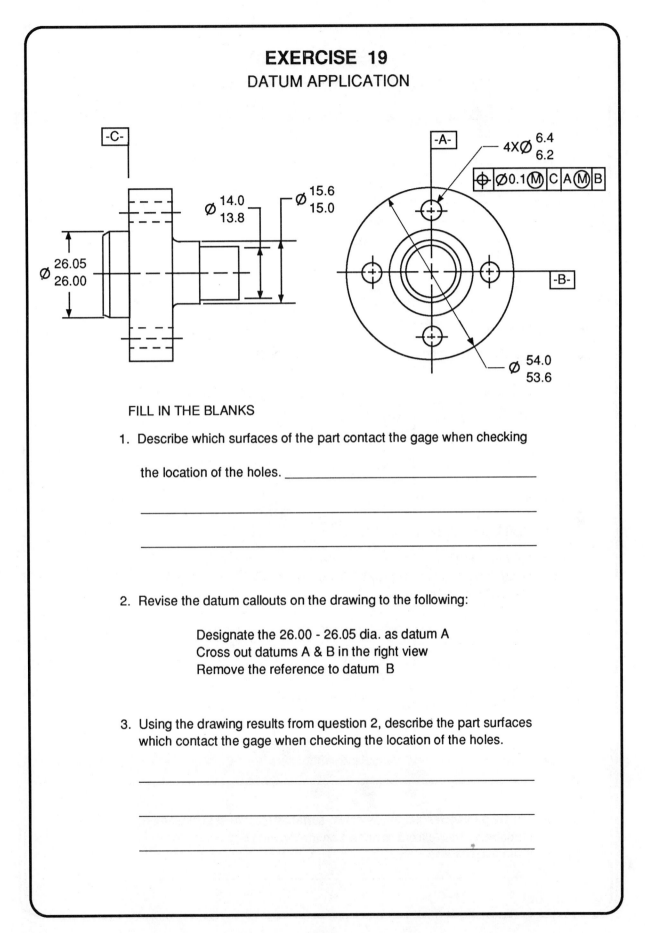

FILL IN THE BLANKS

1. Describe which surfaces of the part contact the gage when checking

   the location of the holes. _____

   _____

   _____

2. Revise the datum callouts on the drawing to the following:

   > Designate the 26.00 - 26.05 dia. as datum A
   > Cross out datums A & B in the right view
   > Remove the reference to datum B

3. Using the drawing results from question 2, describe the part surfaces
   which contact the gage when checking the location of the holes.

   _____

   _____

   _____

EXERCISE 20
PERPENDICULARITY APPLIED TO A FEATURE

1. Is a datum reference always required for an orientation control?

   _____

2. What is the virtual condition of...

   the height of the block? _____

   the width of the block? _____

   the width of the slot? _____

3. What is the flatness limitation of...

   surface "B"? _____

   surface "C"? _____

4. Describe the tolerance zone for surface "C", as established by
   the $\dfrac{29.2}{29.0}$ dimension, as established by the feature control

   symbol. _____

   _____

   _____

5. What controls the squareness of two surfaces (shown at 90°) when no
   perpendicularity callout is shown on the drawing? _____

   _____

220

EXERCISE 21
PERPENDICULARITY APPLIED TO A FEATURE-OF-SIZE

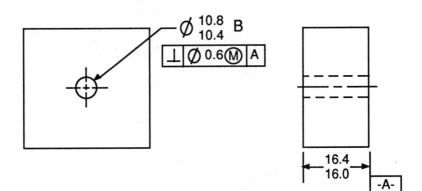

FILL IN THE BLANKS

1. Describe the shape and size of the tolerance zone for the perpendicularity callout of hole B. _____

    _____

2. What is the largest diameter gage pin that would always fit into hole B?

    _____

3. Does the ⊥ callout apply to a feature or a feature-of-size?

    _____

4. What is the straightness of hole B limited to? _____

5. What is the virtual condition of hole B? _____

6. Draw and dimension a gage for checking the orientation of hole B.

EXERCISE 22
ANGULARITY

∠ 0.3 A   SURFACE B

45°

-A-

FILL IN THE BLANKS

1. What is the flatness of surface B limited to? _____

2. Describe the tolerance zone for the angularity callout. _____

   _____

   _____

3. Can angularity be checked with a fixed gage? _____ Why?

   _____

4. Can angularity have a cylindrical tolerance zone? _____

5. Is it permissible to use a toleranced angle to describe a surface which

   is controlled with an angularity callout? _____

   Why? _____

   _____

   _____

EXERCISE 23
PARALLELISM

SURFACE B

// | 0.2 | A

12.4
12.0

- A -

FILL IN THE BLANKS

1. Describe the shape of the tolerance zone for surface B. _____

   _____

2. What is the maximum allowable flatness error on suface B? _____

   _____

   Datum feature A? _____

3. Name three possible tolerance zone shapes for a parallelism control.

   _____

   _____

   _____

4. Does rule #1 apply to the height of the block? _____

5. Is a datum reference always required when using a parallelism control?

   _____

6. What controls the parallelism of two surfaces when no parallelism control is

   shown on the drawing? _____

EXERCISE 24
POSITIONAL TOLERANCING REQUIREMENTS

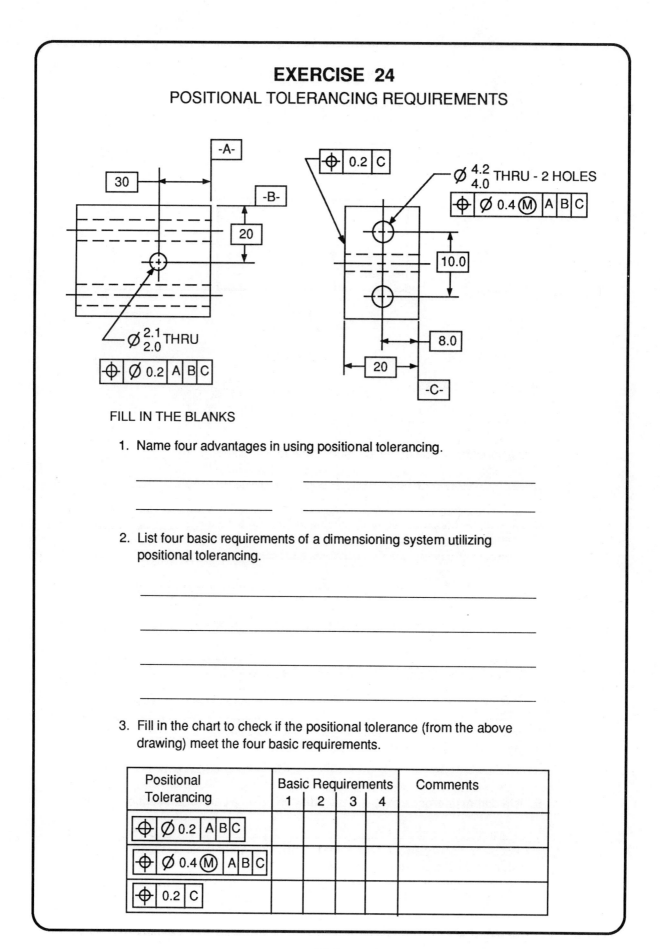

FILL IN THE BLANKS

1. Name four advantages in using positional tolerancing.

   _____   _____

   _____   _____

2. List four basic requirements of a dimensioning system utilizing positional tolerancing.

   _____

   _____

   _____

   _____

3. Fill in the chart to check if the positional tolerance (from the above drawing) meet the four basic requirements.

Positional Tolerancing	Basic Requirements				Comments
	1	2	3	4	
⊕ ∅ 0.2 A B C					
⊕ ∅ 0.4 Ⓜ A B C					
⊕ 0.2 C					

EXERCISE 25
POSITIONAL TOLERANCING AT MMC

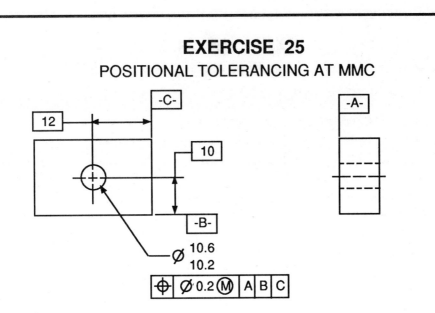

1. Using the drawing above, fill in the chart below.

Hole Size	Positional Tolerance	Bonus	Total Locational Tolerance
10.2			
10.3			
10.4			
10.6			

2. What controls the orientation of the hole relative to datum feature "A"?

_____

3. Fill in the blanks using the words from the list below.

A positional tolerance can be applied to _____.

A fixed gage is used to check a positional tolerance with a _____ modifier.

A positional tolerance zone can be thought of as a _____

control or a _____ control.

RFS	Boundary	A feature-of-size	Axis
MMC	Datum	A feature	Orientation

225

EXERCISE 26
CARTOON GAGES

1. Describe "Cartoon Gage" and its uses. _____

_____

_____

2. List three steps in establishing a cartoon gage.

_____

_____

_____

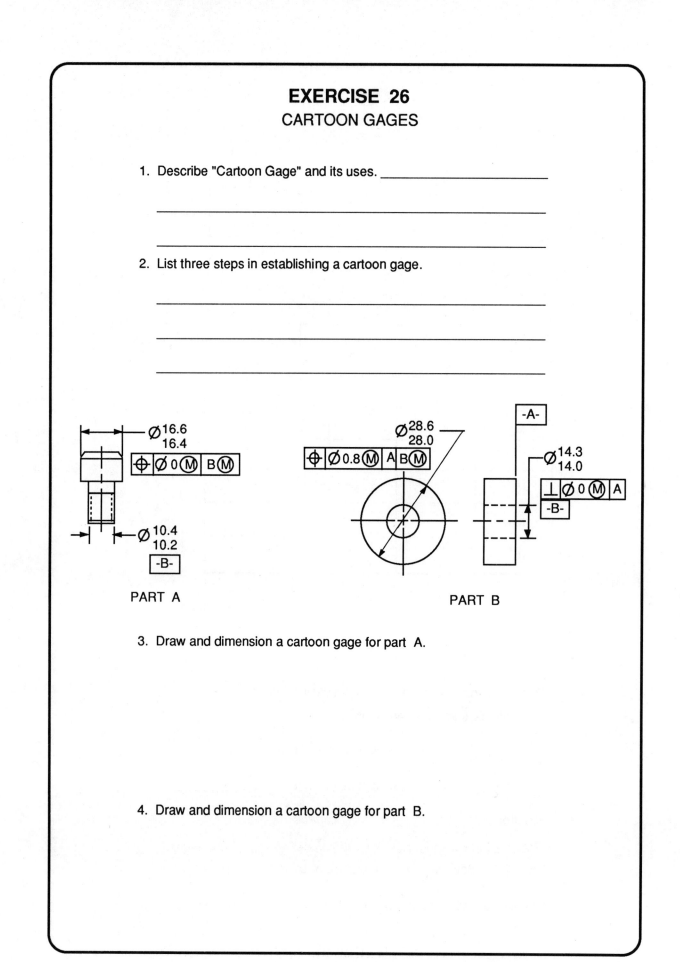

PART A

PART B

3. Draw and dimension a cartoon gage for part A.

4. Draw and dimension a cartoon gage for part B.

EXERCISE 27
CARTOON GAGES

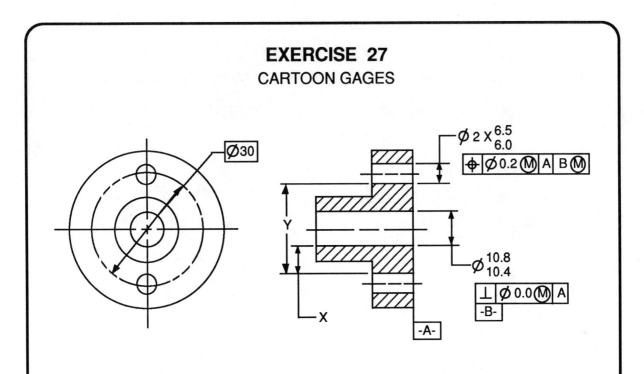

FILL IN THE BLANKS

1. Draw and dimension a cartoon gage for the part above.

2. What is the maximum distance "X" can be?   (use your gage) _____

3. What is the minimum distance "X" can be?   (use your gage) _____

4. What is the maximum distance "Y" can be?   (use your gage) _____

EXERCISE 28
POSITIONAL TOLERANCING COAXIAL FEATURE-OF-SIZE

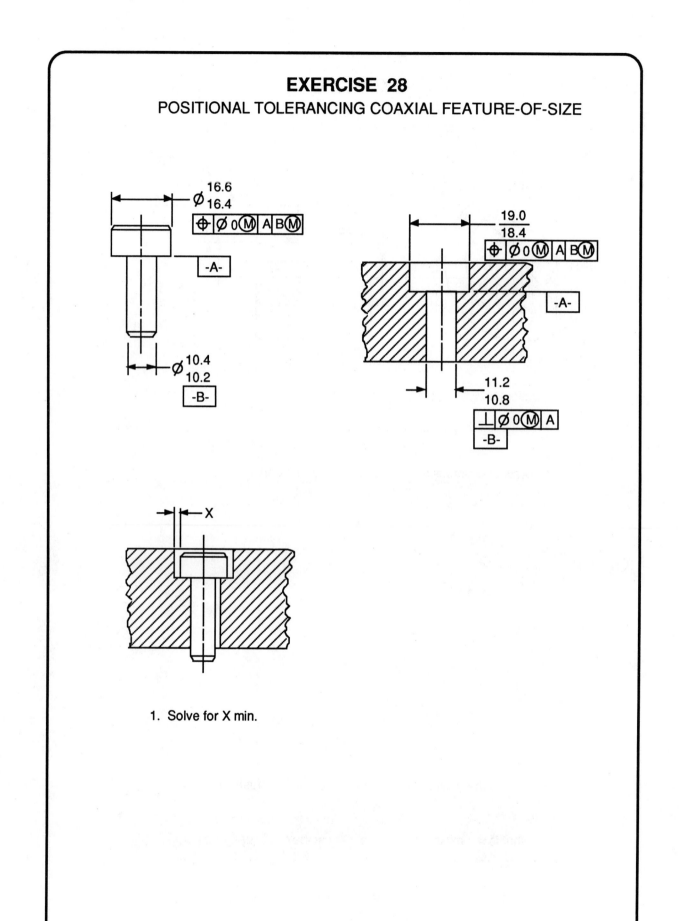

1. Solve for X min.

EXERCISE 29
POSITIONAL TOLERANCING RFS

1. When a positional tolerance is applied RFS, it differs from an MMC application in a number of ways. List three of them.

   _____

   _____

   _____

2. List the most common tolerance zone shapes for positional

   tolerance applications. _____

   _____   _____

3. Use the following description to complete the feature control frame.

   "Regardless of its width, slot A will be within 0.05 of true position relative to datum plane B."

4. What is the shape of the positional tolerance zone in question 3?

   _____

EXERCISE 30
CO-ORDINATE TO POSITIONAL TOLERANCE CONVERSION

USE THE CONVERSION TABLES IN THE APPENDIX TO SOLVE
THE PROBLEMS IN THIS EXERCISE

Convert the co-ordinate tolerances to equivalent round tolerance
zone values.

	$\pm$ x Direction	$\pm$ y Direction		ϕ Tolerance Zone
1.	0.1	0.1	=	_____
2.	0.22	0.12	=	_____
3.	0.16	0.06	=	_____
4.	0.2	0.3	=	_____

5. The print tolerance specifies a hole dia. of 10.1 - 10.6 with a
positional tolerance of 0.25 dia. at MMC. Five parts are checked
and the deviation of the hole from its basic dimension is measured and recorded. The actual hole diameter is also measured
and recorded. The table below shows the recorded data and
provides space for calculated data to be filled out by the inspector
(in this case, you) for each part, calculate the equivalent diameter
tolerance zone, and the bonus tolerance available. Then, combine
the information, compare it to the maximum positional error and
indicate if the part is good or bad.

Part	Deviation x	y	Equiv. ϕ Zone	Actual Hole ϕ	Bonus	Total $\oplus$ Error	Max Permissible $\oplus$ Error	Part Status
1.	0.01	0.02		10.10	0		0.25	GOOD
2.	0.10	0.30		10.32				
3.	0	0.22		10.30				
4.	0.16	0.12		10.14				
5.	0.12	0.28		10.60				

EXERCISE 31
CONCENTRICITY

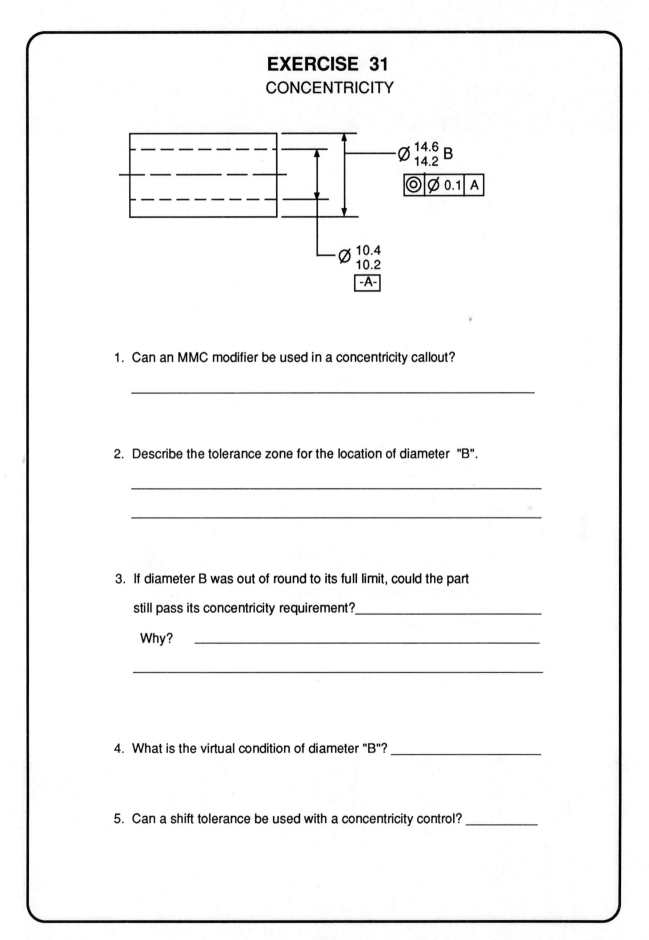

1. Can an MMC modifier be used in a concentricity callout?

   _____

2. Describe the tolerance zone for the location of diameter "B".

   _____

   _____

3. If diameter B was out of round to its full limit, could the part

   still pass its concentricity requirement?_____

   Why?   _____

   _____

4. What is the virtual condition of diameter "B"? _____

5. Can a shift tolerance be used with a concentricity control? _____

EXERCISE 32
RUNOUT

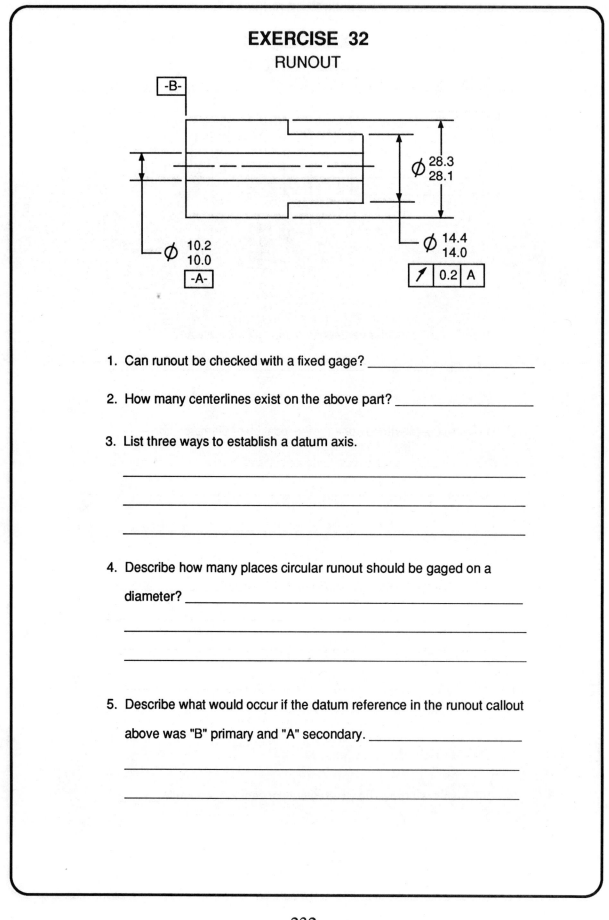

1. Can runout be checked with a fixed gage? _____

2. How many centerlines exist on the above part? _____

3. List three ways to establish a datum axis.

   _____

   _____

   _____

4. Describe how many places circular runout should be gaged on a

   diameter? _____

   _____

   _____

5. Describe what would occur if the datum reference in the runout callout

   above was "B" primary and "A" secondary. _____

   _____

   _____

EXERCISE 33
RUNOUT

Unless otherwise specified all diameters [⟋| 0.2]

---

1. What is the shape of the tolerance zone for circular runout?

_____

_____

2. What is the shape of the tolerance zone for total runout?

_____

_____

3. Is the general runout specification above legal? _____

_____

4. What is the maximum offset between ℄ of diameter "B" and the datum axis? _____

5. What is the maximum offset between ℄ of diameter "D" and the datum axis ? _____

6. What is the virtual condition of diameter ...

B _____

C _____

D _____

EXERCISE 34
CONCENTRICITY/RUNOUT

∅ 20.6 / 20.0

| ↗ | 0.1 | A |

∅ 10.4 / 10.2

-A-

PART A

∅ 20.6 / 20.0

| ◎ | ∅ 0.1 | A |

∅ 10.4 / 10.2

-A-

PART B

1. Describe the size and shape of the tolerance zone for the

   runout callout on part A. _____

   _____

   _____

2. Describe the size and shape of the tolerance zone for the

   concentricity callout on part B? _____

   _____

   _____

3. What is the max possible circularity error on the 20.0 - 20.6 dia.

   of part A? _____ of part B? _____

4. Which part is more expensive to gage A or B? _____

5. Describe what happens if the concentricity symbol is replaced

   with | ⊕ | ∅ 0.1 Ⓢ | A Ⓢ | .

   _____

   _____

   _____

   _____

EXERCISE 35
PROFILE/COORDINATE DIMENSION

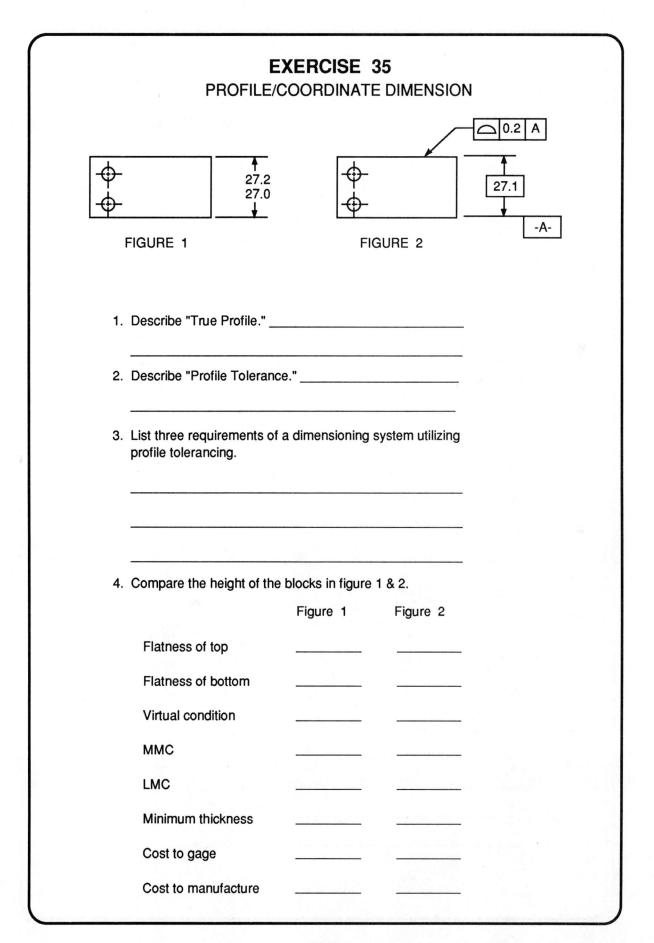

FIGURE 1   FIGURE 2

1. Describe "True Profile." _____

_____

2. Describe "Profile Tolerance." _____

_____

3. List three requirements of a dimensioning system utilizing profile tolerancing.

_____

_____

_____

4. Compare the height of the blocks in figure 1 & 2.

	Figure 1	Figure 2
Flatness of top	_____	_____
Flatness of bottom	_____	_____
Virtual condition	_____	_____
MMC	_____	_____
LMC	_____	_____
Minimum thickness	_____	_____
Cost to gage	_____	_____
Cost to manufacture	_____	_____

EXERCISE 36
PROFILE APPLICATION

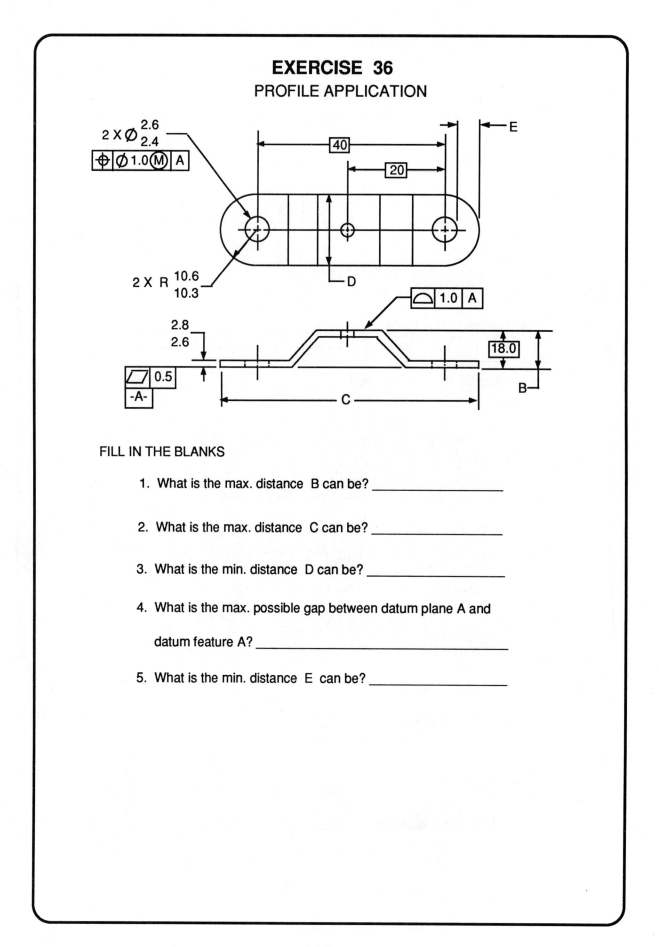

FILL IN THE BLANKS

1. What is the max. distance B can be? _____

2. What is the max. distance C can be? _____

3. What is the min. distance D can be? _____

4. What is the max. possible gap between datum plane A and

   datum feature A? _____

5. What is the min. distance E can be? _____

APPENDIX

CONTENTS

Feature Control Frame Size Proportions	238
Tolerance Zone Conversion	239
Geometric Tolerance Summary Chart	243
Geometric Tolerancing Standards Sources	244
ANSI vs ISO Comparison Chart	245

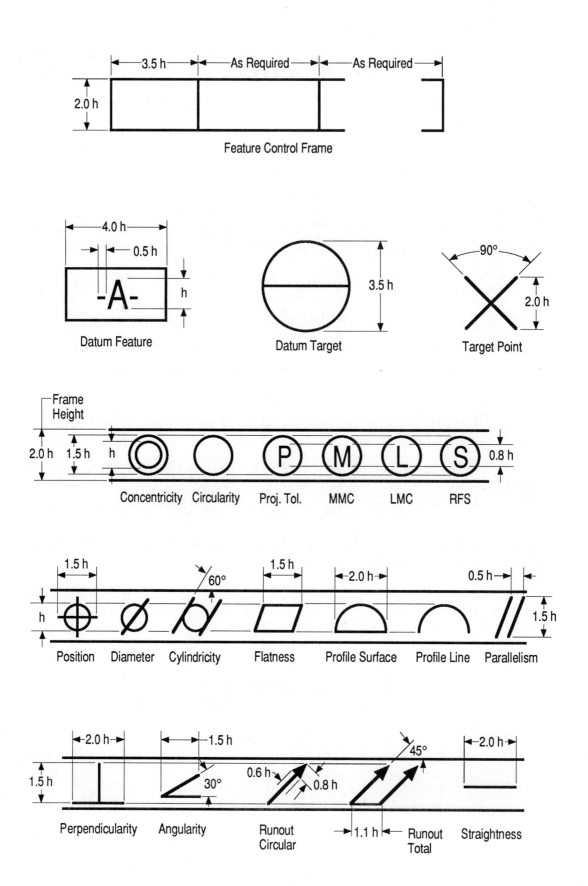

Feature Control Frame

Datum Feature

Datum Target

Target Point

Concentricity   Circularity   Proj. Tol.   MMC   LMC   RFS

Position   Diameter   Cylindricity   Flatness   Profile Surface   Profile Line   Parallelism

Perpendicularity   Angularity   Runout Circular   Runout Total   Straightness

INTRODUCTION

It is often necessary to convert from coordinate tolerances (square or rectangular tolerance zones) to positional tolerances (diametral tolerance zone). The conversion chart in this appendix will help both engineering and manufacturing personnel do this conversion easily.

CONVERSION OF A SQUARE TOLERANCE ZONE TO A DIAMETRAL TOLERANCE ZONE

The square and the circumscribed circle in the lower left hand corner of the conversion chart show the difference between a square and diametral tolerance zone. To use the chart to convert a square tolerance zone to a round tolerance zone do the following:

- Divide the length of the square by 2 to find the equivalent bilateral tolerance.

- Locate the bilateral tolerance value on the X scale of the chart.

- Follow the column up until you reach the underlined number.

- This number represents the equivalent diameter tolerance zone.

Example: A square tolerance zone of 0.32 is equivalent to a 0.16 bilateral tolerance. This converts to a diametral tolerance zone of 0.453.

CONVERSION OF A RECTANGULAR TOLERANCE ZONE TO A DIAMETRAL TOLERANCE ZONE

The conversion chart can also be used to convert a rectangular tolerance zone to a diametral zone. Figure A1 shows the relationship between the two kinds of tolerance zones.

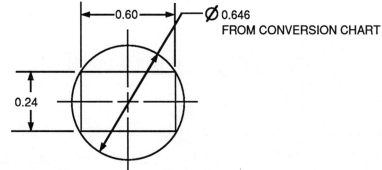

FIGURE A2

239

To use the conversion chart to convert a rectangular tolerance zone to a diametral tolerance zone do the following:

- Convert the rectangular tolerance zone to bilateral tolerances. Divide the length and width by 2.

- Locate both numbers along the bottom and left side of the chart. Draw a vertical and horizontal line into the chart. The number at the intersection point is the equivalent diametral tolerance zone.

Example: Convert the rectangular tolerance zone in figure A2 to an equivalent diametral tolerance zone.

- Divide the length and width by 2: 0.24/2 = 0.12 in the Y direction and 0.60/2 = 0.30 in the X direction.

- Locate 0.12 on the left side and 0.30 along the bottom of the chart. Draw a vertical and horizontal line into the chart. The number at the intersection point, 0.646 is the equivalent diametral zone.

CONVERSION OF COORDINATE MEASUREMENTS TO A DIAMETRAL TOLERANCE ZONE

In figure A3 a common condition for one hole located by basic dimensions from two datums is shown. The location of the hole is inspected by a coordinate measuring machine from the datums specified, and the difference between the actual measurement and the basic dimensions in each direction is noted.

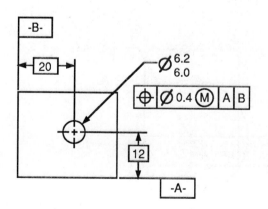

FIGURE A3

In figure A4 the differences are labeled X difference and Y difference. The Z diameter is the actual positional tolerance diameter from the conversion chart. The location of the hole is measured , and the axis is found to be 0.12 in the X direction and 0.24. Should the part be accepted or rejected?

Enter the conversion chart at the 0.12 column on the X scale and move up to the row headed 0.24 on the Y scale. Find the Z value, 0.537. This is the diameter of the tolerance zone. If the print dimensions were RFS, this part would be rejected. However, since the MMC modifier is used in the feature control frame, a bonus tolerance is available and may make the part acceptable.

The actual diameter of the hole must be found to determine how much bonus tolerance is available. The hole is measured and is 6.16 in diameter. It exceeds its MMC by 0.16 so the bonus tolerance available is 0.16. Add the 0.16 bonus to the 0.4 stated positional tolerance and the total permissible tolerance becomes 0.56 diameter. Since the actual hole location is within a 0.537 diameter it is less than the allowable 0.56 diameter limit, and the part is acceptable.

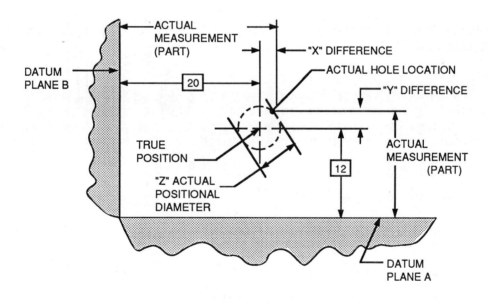

FIGURE A4

CONVERSION CHART

Coordinate Measurement to Diameter Tolerance Zone

DIAMETER ZONE (Z)

Y \ X	0.02	0.04	0.06	0.08	0.10	0.12	0.14	0.16	0.18	0.20	0.22	0.24	0.26	0.28	0.30	0.32	0.34	0.36	0.38	0.40	0.42	0.44	0.46	0.48	0.50
0.50	1.001	1.003	1.007	1.013	1.020	1.028	1.038	1.050	1.063	1.077	1.093	1.109	1.127	1.146	1.166	1.187	1.209	1.232	1.256	1.281	1.306	1.332	1.359	1.386	1.414
0.48	0.961	0.963	0.967	0.973	0.981	0.990	1.000	1.012	1.025	1.040	1.056	1.073	1.092	1.111	1.132	1.154	1.176	1.200	1.224	1.250	1.276	1.302	1.330	1.358	1.386
0.46	0.921	0.923	0.928	0.934	0.941	0.951	0.962	0.974	0.988	1.003	1.020	1.038	1.057	1.077	1.098	1.121	1.144	1.168	1.193	1.219	1.246	1.273	1.301	1.330	1.359
0.44	0.881	0.884	0.888	0.894	0.902	0.912	0.923	0.936	0.951	0.967	0.984	1.002	1.022	1.043	1.065	1.088	1.112	1.137	1.163	1.189	1.217	1.245	1.273	1.302	1.332
0.42	0.841	0.844	0.849	0.855	0.863	0.874	0.885	0.899	0.914	0.930	0.948	0.967	0.988	1.010	1.032	1.056	1.081	1.106	1.133	1.160	1.188	1.217	1.246	1.276	1.306
0.40	0.801	0.804	0.809	0.816	0.825	0.835	0.848	0.862	0.877	0.894	0.913	0.933	0.954	0.977	1.000	1.024	1.050	1.076	1.103	1.131	1.160	1.189	1.219	1.250	1.281
0.38	0.761	0.764	0.769	0.777	0.786	0.797	0.810	0.825	0.841	0.859	0.878	0.899	0.921	0.944	0.968	0.994	1.020	1.047	1.075	1.103	1.133	1.163	1.193	1.224	1.256
0.36	0.721	0.724	0.730	0.738	0.747	0.759	0.773	0.788	0.805	0.824	0.844	0.865	0.888	0.912	0.937	0.963	0.990	1.018	1.047	1.076	1.106	1.137	1.168	1.200	1.232
0.34	0.681	0.685	0.691	0.699	0.709	0.721	0.735	0.752	0.769	0.789	0.810	0.832	0.856	0.881	0.907	0.934	0.962	0.990	1.020	1.050	1.081	1.112	1.144	1.176	1.209
0.32	0.641	0.645	0.651	0.660	0.671	0.684	0.699	0.716	0.734	0.755	0.777	0.800	0.825	0.850	0.877	0.905	0.934	0.963	0.994	1.024	1.056	1.088	1.121	1.154	1.187
0.30	0.601	0.605	0.612	0.621	0.632	0.646	0.662	0.680	0.700	0.721	0.744	0.768	0.794	0.821	0.849	0.877	0.907	0.937	0.968	1.000	1.032	1.065	1.098	1.132	1.166
0.28	0.561	0.566	0.573	0.582	0.595	0.609	0.626	0.645	0.666	0.688	0.712	0.738	0.764	0.792	0.821	0.850	0.881	0.912	0.944	0.977	1.010	1.043	1.077	1.111	1.146
0.26	0.522	0.526	0.534	0.544	0.557	0.573	0.591	0.611	0.632	0.656	0.681	0.708	0.735	0.764	0.794	0.825	0.856	0.888	0.921	0.954	0.988	1.022	1.057	1.092	1.127
0.24	0.482	0.487	0.495	0.506	0.520	0.537	0.556	0.577	0.600	0.625	0.651	0.679	0.708	0.738	0.768	0.800	0.832	0.865	0.899	0.933	0.967	1.002	1.038	1.073	1.109
0.22	0.442	0.447	0.456	0.468	0.483	0.501	0.522	0.544	0.569	0.595	0.622	0.651	0.681	0.712	0.744	0.777	0.810	0.844	0.878	0.913	0.948	0.984	1.020	1.056	1.093
0.20	0.402	0.408	0.418	0.431	0.447	0.466	0.488	0.512	0.538	0.566	0.595	0.625	0.656	0.688	0.721	0.755	0.789	0.824	0.859	0.894	0.930	0.967	1.003	1.040	1.077
0.18	0.362	0.369	0.379	0.394	0.412	0.433	0.456	0.482	0.509	0.538	0.569	0.600	0.632	0.666	0.700	0.734	0.769	0.805	0.841	0.877	0.914	0.951	0.988	1.025	1.063
0.16	0.322	0.330	0.342	0.358	0.377	0.400	0.425	0.453	0.482	0.512	0.544	0.577	0.611	0.645	0.680	0.716	0.752	0.788	0.825	0.862	0.899	0.936	0.974	1.012	1.050
0.14	0.283	0.291	0.305	0.322	0.344	0.369	0.396	0.425	0.456	0.488	0.522	0.556	0.591	0.626	0.662	0.699	0.735	0.773	0.810	0.848	0.885	0.923	0.962	1.000	1.038
0.12	0.243	0.253	0.268	0.288	0.312	0.339	0.369	0.400	0.433	0.466	0.501	0.537	0.573	0.609	0.646	0.684	0.721	0.759	0.797	0.835	0.874	0.912	0.951	0.990	1.028
0.10	0.204	0.215	0.233	0.256	0.283	0.312	0.344	0.377	0.412	0.447	0.483	0.520	0.557	0.595	0.632	0.671	0.709	0.747	0.786	0.825	0.863	0.902	0.941	0.981	1.020
0.08	0.165	0.179	0.200	0.226	0.256	0.288	0.322	0.358	0.394	0.431	0.468	0.506	0.544	0.582	0.621	0.660	0.699	0.738	0.777	0.816	0.855	0.894	0.934	0.973	1.013
0.06	0.126	0.144	0.170	0.200	0.233	0.268	0.305	0.342	0.379	0.418	0.456	0.495	0.534	0.573	0.612	0.651	0.691	0.730	0.769	0.809	0.849	0.888	0.928	0.967	1.007
0.04	0.089	0.113	0.144	0.179	0.215	0.253	0.291	0.330	0.369	0.408	0.447	0.487	0.526	0.566	0.605	0.645	0.685	0.724	0.764	0.804	0.844	0.884	0.923	0.963	1.003
0.02	0.057	0.089	0.126	0.165	0.204	0.243	0.283	0.322	0.362	0.402	0.442	0.482	0.522	0.561	0.601	0.641	0.681	0.721	0.761	0.801	0.841	0.881	0.921	0.961	1.001

ROUND TOLERANCE ZONE

ØZ

SQUARE COORDINATE ZONE

FORMULA $Z = 2\sqrt{x^2 + y^2}$

GEOMETRIC TOLERANCING REFERENCE CHART (PER ANSI Y14.5 M-1982)

TYPE OF TOLERANCE	GEOMETRIC CHARACTERISTIC	SYMBOL	CAN BE APPLIED TO A FEATURE	CAN BE APPLIED TO A FEATURE OF SIZE	CAN AFFECT VIRTUAL CONDITION	DATUM REFERENCE USED?	CAN USE Ⓜ MODIFIER	CAN USE Ⓢ MODIFIER	CAN BE AFFECTED BY A BONUS TOLERANCE	CAN BE AFFECTED BY A SHIFT TOLERANCE
FORM	STRAIGHTNESS	—	YES	YES	YES*	NO	YES*	NO•	YES □	NO
FORM	FLATNESS	▱	YES	NO	NO	NO	NO	NO•	NO	NO
FORM	CIRCULARITY	○	YES	NO	NO	NO	NO	NO•	NO	NO
FORM	CYLINDRICITY	⌭	YES	NO	NO	NO	NO	NO•	NO	NO
ORIENTATION	PERPENDICULARITY	⊥	YES	YES	YES*	YES	YES*	NO•	YES □	YES ○
ORIENTATION	ANGULARITY	∠	YES	YES	YES*	YES	YES*	NO•	YES □	YES ○
ORIENTATION	PARALLELISM	∥	YES	YES	YES*	YES	YES*	NO•	YES □	YES ○
LOCATION	POSITIONAL TOLERANCE	⌖	NO	YES	YES	YES	YES	YES	YES □	YES ○
LOCATION	CONCENTRICITY	◎	YES	YES	YES*	YES	NO	YES	NO	NO
RUNOUT	CIRCULAR RUNOUT	↗	YES	NO	NO	YES	NO	NO•	NO	NO
RUNOUT	TOTAL RUNOUT	⌰	YES	NO	NO	YES	NO	NO•	NO	NO
PROFILE	PROFILE OF A LINE	⌒	YES	NO	NO	YES**	NO	NO•	NO	YES ○
PROFILE	PROFILE OF A SURFACE	⌓	YES	NO	NO	YES**	NO	NO•	NO	YES ○

* WHEN APPLIED TO A FEATURE-OF-SIZE

** CAN ALSO BE USED AS A FORM CONTROL WITHOUT A DATUM REFERENCE

○ WHEN A DATUM FEATURE-OF-SIZE IS REFERENCED WITH THE MMC MODIFIER.

□ WHEN AN MMC MODIFIER IS USED.

• AUTOMATIC PER RULE #3

ORDERING STANDARDS

To Order GM Drafting Standards:

General Motors Corporation
CPE - Engineering Standards - N2
Engineering Building
GM Technical Center
30200 Mound Road
Warren, Michigan   48090-9010

To Order ANSI Y14.5M - 1982 Dimensioning and Tolerancing Standards:

American Society of Mechanical Engineers
United Engineering Center
345 East 47th Street
New York, N.Y. 10017

COMPARISON ANSI Y14.5, AND ISO SYMBOLS

CHARACTERISTIC	ANSI - Y14.5	ISO 1101
STRAIGHTNESS	—	—
FLATNESS	▱	▱
ANGULARITY	∠	∠
PERPENDICULARITY (SQUARENESS)	⊥	⊥
PARALLELISM	//	//
CONCENTRICITY	◎	◎
POSITION	⊕	⊕
CIRCULARITY (ROUNDNESS)	○	○
SYMMETRY	⊕	≡
PROFILE OF A LINE	⌒	⌒
PROFILE OF A SURFACE	◠	◠
ALL AROUND - PROFILE	⌀—	NONE
RUNOUT (CIRCULAR)	↗	↗
RUNOUT (TOTAL)	↗↗	↗↗
CYLINDRICITY	⌀	⌀
DATUM FEATURE	-A-	⏚ OR ⏚ A
MAXIMUM MATERIAL CONDITION (MMC)	Ⓜ	Ⓜ
REGARDLESS OF FEATURE SIZE (RFS)	Ⓢ	NONE (ASSUMED UNLESS SPECIFIED MMC)
LEAST MATERIAL CONDITION (LMC)	Ⓛ	NONE
BASIC DIMENSION	XX	XX

ANSWER GUIDE

ANSWER SHEET CHAPTER 1 PROBLEMS (ODD)

1. Communications tool, design philosophy

3. Improves communications
   Better design decisions
   Production tolerances increased

5. Maximum material condition
   Least material condition
   Regardless of feature size
   Projected tolerance zone
   Diameter

7. The theoretical extreme boundry of a feature-of-size generated by the collective effects of the MMC, and any applicable geometric tolerances.

9. A bonus tolerance is an extra tolerance which is available whenever an MMC modifier is used in the tolerance portion of a feature control frame.

11.

LETTER	FEATURE	FEATURE OF SIZE	LOCATION DIMENSION
A		X	
B	X		
C	X		
D	X		
E		X	
F	X		
G		X	
H			X

13.

LETTER	MMC	LMC
A	20.0	19.0
B	4.9	5.1
C	7.8	7.2
D	21.0	20.6
E	6.2	6.8

15. 20.4

17.

ACTUAL HOLE Ø	THE BONUS TOL AVAILABLE IS ...	
10.4	0.4	(RFS IMPLIED)
10.3	0.3	
10.2	0.2	
10.1	0.1	
10.0	0	
9.8	BAD PART	(OUT OF SPEC CAN BE REWORKED)

19.

ACTUAL HOLE Ø	THE BONUS TOL. AVAILABLE IS...
11.0	0
10.8	0.2
10.6	0.4
10.4	0.6
10.2	0.8
10.0	1.0

Crossword Puzzle Answers...

```
      MILK LOW  DATUM
      U  RIM    E    SCRAP
      S  FUNCTIONAL
M     C  L      E S        F
ORAL  VIGOR  MMC   THREE
D     E  K        O   O    A
I     FORM  VIRTUAL       T
F     A  W  E          E  U
I     BONUS  N          PRE
E     A  S K DESIGNER     E
RFS      I   I    U    A
      I      IMPROVES    COG
ICE   P      F      Y  F   E
T     BOSS   A         L   O
U     S  S      CIRCULAR   M
R     T  I  RULE  E     T  E
BLUE     E         L VAN T
I     PHILOSOPHY      EVER
NOW      O      A      S   I
E     ANGULARITY       S   C
```

249

ANSWER SHEET CHAPTER 2 PROBLEMS (ODD)

1. 0.1

3. No

5. Less

7. Zero

9. 36.4

11. 0.2

13. Rule #1

15. 10.5

17. Zero

19. Useless control, Rule #1 provides a tighter control.

21. Zero

23. 0.04

25. No, Feature control is applied to a feature-of-size, and applies to part axis.

27. 0.80

29. No, it is overridden.

31. 12.5

33. Zero

35. 0.24999...

37. No

39.

IF THE SYMBOLOGY AT LOCATION #1 WAS...	MAX ALLOWABLE FLATNESS ERROR ON SURFACE ...		MIN SETTING OF PARALLEL PLATES WHICH THE PART WILL PASS THRU	VIRTUAL CONDITION OF PLATE THICKNESS
	B	C		
SAME AS SPECIFIED	0.1	0.16	4.97	4.97
REVISED TO READ $\boxed{\diagup\!\!\diagup \;\; 0.1 \;\; C}$	ILLEGAL SYMBOL	0.16	4.97	4.97
REMOVED	0.16	0.16	4.97	4.97

Crossword Puzzle Answers...

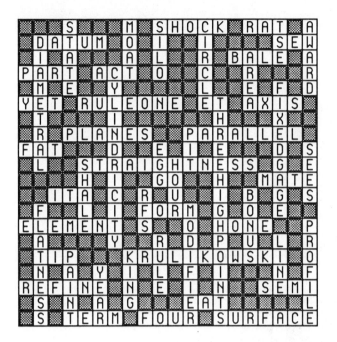

ANSWER SHEET CHAPTER 3 PROBLEMS (ODD)

1. ??? Can't do - datum sequence is not designated.

3. $\boxed{90°}$

5. 3 points for "A", and 2 points for "B."

7. It is a feature-of-size. There is no geometric tolerance relating it to the datum reference frame.

9. Line

11. No, because the 3-2-1 rule only applies with all planar datums.

13. Gage, drawing

15. Lines

17. Datum "A"

19. Yes

21. No, they are considered gage dimensions.

23. No

25. Yes

Crossword Puzzle Answers...

```
B  SECONDARY    RARE  B
LIE O   O A       U R I
A   L   O  PLOTS   RARE
S   FEATURE    TAX   T H
T   T    X    P    R  H D
  S  U P A  CAB  E    D
 TERM O C   R  P    L A
  I  SHIFT   TERTIARY
FRY   N       I   K
    DATUM     M  READ
BIN  I    B   A     C
 D DATUMFEATURE   TON
 L  M   M   S  Y    S
 ELDER  C  LINE F    M
  O  T  P   C NOON S A
COT E LAD    D R  STEP P
A  AREA   G      I R
BEE    N PARTFUNCTION
L  DATA  G   A    D A
E     REFERENCE   E P
```

ANSWER SHEET CHAPTER 4 PROBLEMS (ODD)

1. 2 parallel planes 0.05 apart

3. No

5. 0.05

7. RFS, Rule #3

9. A cylinder 0.1 diameter

11. No, because hole V.C. = 10.2 - 0.1 = 10.1

13.

ACTUAL FEATURE SIZE (DIA)	TOLERANCE ZONE DIAMETER
6.6	0.3
6.5	0.4
6.4	0.5
6.3	0.6
6.2	0.7
6.1	BAD PART

15.

ACTUAL FEATURE SIZE	WIDTH OF PERPENDICULARITY TOLERANCE ZONE
24.7	0.15
24.8	0.15
24.9	0.15
25.0	0.15
25.1	0.15
25.2	0.15
25.3	0.15

17. Two parallel planes, parallel with each other and parallel to datum plane "A", spaced 0.2 apart, located within the size limits, within which the surface must lie.

19. 36.5

21. Relative to each other - the size dimension and Rule #1.
    Relative to datum plane "A" - the location dimension, the size dimension and Rule #1

23. 0.3

25. ?? Uninterpretable - A basic angle is required relative to datum plane "B".

Crossword Puzzle Answers...

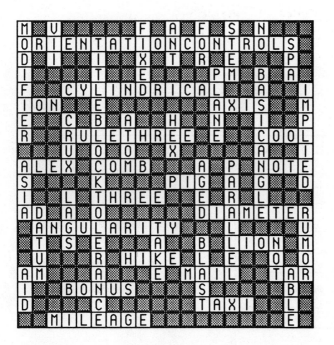

ANSWER SHEET CHAPTER 5 PROBLEMS (ODD)

1. A

3. 5.3

5. 14.0 from "B", 10.0 from "C"

7. 0.4

9. 10.85

11.

-B-

14.0

Ø 5.3 POST

10.0

-C-

SURFACE OF PAGE REPRESENTS DATUM PLANE "A"

13. Cylindrical

15. 0.3 @ MMC, 0.9 @ LMC

17.

POST SIZE	POSITIONAL TOLERANCE	BONUS TOLERANCE	TOTAL LOCATION TOLERANCE
8.6	0.3	0.0	0.3
8.5	0.3	0.1	0.4
8.4	0.3	0.2	0.5
8.3	0.3	0.3	0.6
8.2	0.3	0.4	0.7
8.1	0.3	0.5	0.8
8.0	0.3	0.6	0.9

19. 8.0

21. 0.2

23.

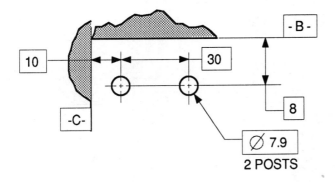

SURFACE OF PAGE REPRESENTS
DATUM PLANE "A"

25. 21.5

27. 0.2

29. 6.8

31. 0.6 @ MMC, 0.8 @ LMC.

33. 10.65, 9.15

35.

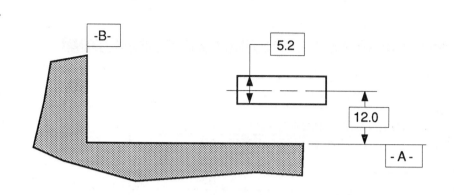

37. 9.4

39. 1.0

41. 18.0

43.

DISTANCE	MAXIMUM	MINIMUM
X	17.0	12.0
Y	52.0	46.0
Z	*	*

* INSUFFICIENT DATA
  TO CALCULATE

Crossword Puzzle Answers...

ANSWER SHEET CHAPTER 6 PROBLEMS (ODD)

1.   A single diameter.
     Two diameters a sufficient distance apart.
     A surface primary and a diameter secondary.

3.   a. Location
     b. Circularity

5.   FALSE - The part is rotated about the datum axis.

7.

	MAXIMUM	MINIMUM
D	3.45	2.25
E	6.90	5.1
F	3.55	2.95
G	6.80	5.40
H	9.35	8.35

9.   No

11. An 11.8 diameter pin, because the hole V.C. is 12.0 - 0.2 = 11.8

Crossword Puzzle Answers...

ANSWER SHEET CHAPTER 7 PROBLEMS (ODD)

1. Two parallel planes 0.2 apart.

3. Unilateral

5. No

7. 24.0

9. Gaging of the part would not be repeatable.

11. Datums "A" and "B" are co-primary datums.

13. 24.3

15. 9.3

17. 38.2

19. A combination of Rule #1 and the size dimension.

Crossword Puzzle Answers...

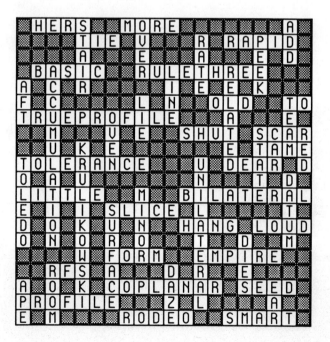

INDEX

A

Angularity
      Applications; 106, 107
      Legal Specification Test; 108
      Tolerance Zones; 96, 106, 107

B

Basic Dimension
      Angularity; 106
      As Gage Dimensions; 22
      Cartoon Gages; 132, 133
      Datum Planes; 70
      Definition; 22
      Implied Zero; 144
      Locating Datum Targets; 74
      Positional Tolerance; 123
      Profile Tolerance; 178, 179, 182, 184, 187
Bonus Tolerance
      Definition; 23
      Parallelism; 111
      Perpendicularity; 101, 102
      Positional Tolerance; 123, 130, 131, 134, 135
      Straightness; 46, 47

C

Cartoon Gage; 132, 133
Circular Runout; (See Runout Circular)
Circularity
      Application; 51
      Circular Runout; 163
      Definition; 49
      Indirect Controls; 52
      Legal Specification Test; 52
      Rule #1; 50
      Symbol; 11
      Total Runout; 168
Coaxiality
      Application; 144, 161, 163, 165, 166, 169
      Comparison; 170
      Definition; 144
Concentricity
      Application; 149
      Definition; 148
      Symbol; 11

Coplanar Surfaces; (See Profile Of A Surface)
Cylindricity
    Application; 54
    Definition; 53
    Indirect Controls; 55
    Legal Specification Test; 56
    Rule #1; 54
    Symbol; 11

D

Datum
    3-2-1 Rule; 72
    Applications (MMC); 82
    Co-Datums; 85
    Datum Axis; 162
    Datum Feature; 66
    Datum Precedence; 71
    Datum Reference Frame; 70
    Datum Shift; 84
    Definition; 66
    Feature-Of-Size; 77, 86
    Planar Feature Datums; 69
    Purpose Of; 66
    Selection; 68
    Specifying; 66, 67
    Virtual Condition Datums; 82
Datum Targets
    Applications; 74
    Area; 76
    Definition; 73
    Line; 75
    Point; 74
    Symbol; 73
Diameter Symbol; 12

F

Feature
    Definition; 8
Feature-Of-Size
    Definition; 8
Feature Control Frame
    Application; 14
    Definition; 13
    Examples; 13

Flatness
    Application; 37
    Definition; 35
    Indirect Flatness Controls; 37
    Legal Specification Test; 38
    Rule #1; 36
    Symbol; 11
Functional Dimensioning; 7
Functional Gage; 130, 134
Fundamental Rules; (See Rules)

G

Geometric Characteristic Symbols; 11
Geometric Dimensioning And Tolerancing
    Advantages; 6
    Definition; 5
    Disadvantages; 6

H

History of Geometric Dimensioning And Tolerancing; 2

L

Least Material Condition
    Application; 140
    Definition; 10
    Symbol; 12
Location Controls; 122
Location Dimension; 8

M

Maximum Material Condition
    Application; 23
    Definition; 9
    Symbol; 12
Modifiers; 12

O

Orientation Controls
    Definition; 96
    General Information; 96
    Tolerance Zones; 96
    (See Also Angularity, Perpendicularity, And Parallelism)

P

Parallelism
    Applied To A Surface; 109
    Definition; 109
    Legal Specification Test; 113
    Of An Axis; 111
    Symbol; 11
    Using MMC Modifier; 111
Perpendicularity
    Applied To An Axis; 100, 101
    Applied To A Centerplane; 99
    Applied To A Surface; 98
    Applied To A Surface Element; 103
    General Cases; 97
    Legal Specification Test; 105
    Symbol; 11
    Tolerance; 97
Positional Tolerance
    Advantages; 123
    Applications (Coaxial Features-Of-Size); 144, 145
    Applications (LMC); 140
    Applications (MMC); 130, 131, 132, 133, 134, 136, 137
    Applications (Planar Feature-Of-Size); 141
    Applications (RFS); 138, 139
    Applications (Symmetry); 146, 147
    Axis Concept; 128, 129
    Bonus Tolerance; 123, 130, 131
    Boundary Concept; 126, 127
    Definitions; 123
    Gaging; 130, 132, 133, 134
    General Information; 122
    Requirements; 124
    Symbol; 11
Profile
    Advantages; 178
    General Information; 178
    Tolerance Zones; 179
Profile Of A Line
    Application; 186
    Legal Specification Test; 187, 188
    Symbol; 11
    Tolerance Zone; 186
Profile Of A Surface
    Applications; 180, 181, 182, 183
    Legal Specification Test; 184, 185
    Symbol; 11
    Tolerance Zone; 180

Projected Tolerance Zone
    Definition; 142
    Symbol; 12

R

Regardless Of Feature Size
    Definition; 10
    Symbol; 12
Rules
    Rule #1; 15, 16, 17, 18
    Rule #2; 19
    Rule #3; 20
Runout
    Composite Control; 160
    Establishing A Datum Axis; 162
    Example; 161
    Features Applicable To Runout; 161
    General Information; 160
    Part Calculations; 171, 172
Runout Circular
    Composite Control; 160
    Definition; 163, 164
    Legal Specification Test; 167
    Surface Application; 166
    Symbol; 11, 163
    Tolerance Zone; 165
Runout Total
    Definition; 168
    Legal Specification Test; 170
    Surface Application; 169
    Symbol; 11, 160
    Tolerance Zone; 168, 169

S

Shift Tolerance
    Definition; 84
    Tolerance Of Position; 135
Straightness Of A Feature
    Application; 41
    Definition; 39
    Legal Specification Test; 42
    Rule #1; 40
    Symbol; 11
    Tolerance Zone; 39

Straightness Of Features-Of-Size
    Application (MMC); 46
    Application (RFS); 45
    Definition; 43
    Legal Specification Test; 48
    Rule #1; 44
Symbols; 11
Symmetry (See Positional Tolerance)

T

Target, Datum (See Datum Target)
Tolerance (See Bonus Tolerance)

V

Virtual Condition
    Datums; 82, 83
    Definition; 24, 25